THE MARINE SEISMIC SOURCE

SEISMOLOGY AND EXPLORATION GEOPHYSICS

GREGG PARKES

and

LES HATTON

Merlin Profilers (Research) Ltd., U.K.

THE MARINE SEISMIC SOURCE

SPRINGER-SCIENCE+BUSINESS MEDIA, B.V.

Library of Congress Cataloging in Publication Data

CIP

Parkes, Gregg, 1954-
The marine seismic source.

(Seismology and exploration geophysics)
Bibliography : p.
Includes index.
1. Submarine geology. 2. Seismology–Data processing. 3. Seismic reflection method–Deconvolution. I. Hatton, Les, 1948- II. Title. III. Series.
QE39.P37 1986 551.2'2 86-3278
ISBN 978-90-481-8425-5 ISBN 978-94-017-3385-4 (eBook)
DOI 10.1007/978-94-017-3385-4

TABLE OF CONTENTS

Preface vii
Acknowledgements ix

1. Underlying Physics and Concepts 1

1.1. ACOUSTIC WAVE GENERATION TECHNIQUES 3
1.2. ACOUSTIC WAVE PROPAGATION IN WATER 4
1.3. THE RELATIONSHIP BETWEEN VELOCITY, PRESSURE AND ENERGY 7
1.4. BUBBLE MOTION 8
1.5. THE GHOST OR VIRTUAL IMAGE 10
1.6. THE NEAR AND FAR RADIATION FIELDS 15
1.7. THE MEASUREMENT OF SIGNATURES 16
1.8. RELATIVE MOTION EFFECTS 17
1.9. SIGNATURE PARAMETERS 19
1.10. FACTORS DETERMINING CHOICE OF SOURCE 20

2. Source Arrays and Directivity 23

2.1. THE REASONS FOR USING ARRAYS 25
2.2. PULSE SHAPING 25
2.3. THE PRINCIPLES OF DIRECTIVITY 26
2.4. DIRECTIVITY DISPLAYS 29
2.5. BEAM STEERING 31
2.6. DIRECTIVITY OF LONG ARRAYS 33
2.7. DIRECTIVITY OF WIDE ARRAYS 34
2.8. THE USE OF DIRECTIVITY FOR NOISE SUPPRESSION 36
2.9. EFFECTS OF DIRECTIVITY ON PRIMARY DATA 39

3. Interaction and Wavefield Determination 43

3.1. CATEGORIES OF SOURCE INTERACTION 45
3.2. THE WAVEFIELD OF A LINEARLY INTERACTING ARRAY 50
3.3. ENERGY CONSIDERATIONS 59
3.4. THE AVOIDANCE OF INTERACTION EFFECTS 61
3.5. SUMMARY 62

4. Practical Aspects of Wavefield Stability 65

4.1. TIME SYNCHRONISATION 67
4.2. SOURCE GEOMETRY STABILITY 68
4.3. SYSTEM STABILITY 70
4.4. THE EFFECTS OF WEATHER 72

4.5. MARINE GHOST VARIABILITY 74
4.6. SUMMARY 77

5. Source Signature Deconvolution 79

5.1. INTRODUCTION 81
5.2. DETERMINISTIC DECONVOLUTION 83
5.3. STATISTICAL DECONVOLUTION 93
5.4. PRACTICAL NOTES 94

Appendix - Technical Description of Main Sources 97

References 105
Index 111

PREFACE

This book is about marine seismic sources, their history, their physical principles and their deconvolution. It is particularly accented towards the physical aspects rather than the mathematical principles of signature generation in water as it is these aspects which the authors have found to be somewhat neglected. A huge amount of research has been carried out by both commercial and academic institutions over the years and the resulting literature is a little daunting, to say the least. In spite of this, the subject is intrinsically very simple and relies on a very few fundamental physical principles, a somewhat larger number of heuristic principles and a refreshingly small amount of blunderbuss mathematics. As such it is still one of those subjects in which the gifted practical engineer reigns supreme and from which many of the important advances have originated.

In Chapter 1 of the book, the underlying physics and concepts are discussed, including pressure and wave propagation, bubble motion, virtual images and the factors determining choice of source. In marine reflection seismology, almost all of the seismic data acquired currently is done with either the airgun or the watergun, which rely on the expulsion of air and water respectively to generate acoustic energy. As a consequence, the discussion in this chapter is geared towards these two sources, as is much of the rest of the book. Many noteworthy sources have appeared in the past however, which used different physical principles but which nevertheless obey the same basic laws, and these sources are not enlarged upon here as they are now very much in the minority.

Chapter 2 discusses directivity, a much misunderstood phenomenon. The underlying principles are described, as are the directional dependence of radiation for arrays of sources and the methods of exploiting this directivity to improve the quality of seismic data. Since all known sources are directional to some extent as a result of the virtual image or ghost reflection in the sea surface, an understanding of this basically simple subject is important.

The deconvolution of seismic sources is an important topic of this book, as the real goal of marine seismic source studies is to produce something that can promptly and simply be removed by the processing geophysicist ! Such deconvolution techniques break naturally into two categories, the statistical and the deterministic. Deterministic methods require a prior knowledge of the source signature. In practical arrays, such knowledge is available only if the phenomenon of interaction is understood. The problem of interaction arises because any source fired simultaneously with other sources does not behave the same way as when fired in isolation. This rather confusing effect is studied in considerable detail in Chapter 3, followed by a number of proposed methods for determining the source signature, some of which are outstandingly successful. In addition, the fundamental difference between directivity and interaction is also re-emphasized.

The science of marine seismic sources is an eminently pragmatic subject and Chapter 4 goes into considerable detail on the practical aspects of the stability of the radiated acoustic energy field. The effects of source synchronization, source geometry, weather and the surface ghost response are all analysed.

Chapter 5 discusses how all this accumulated knowledge is used to shape the predicted or estimated source radiation field into some ideal response from the point of view of the processing geophysicist. After studying the effects of both simple band-limiting and absorptive band-limiting on the source signature, deterministic deconvolution as well as a number of statistical deconvolution techniques are described, including their advantages and disadvantages from a practical point of view.

Finally, technical descriptions of the most common sources are included as an appendix for reference purposes.

ACKNOWLEDGEMENTS

This book was written after three exhilarating and fairly intense years research into marine seismic sources and most notably the airgun. During this period, many people influenced our understanding but most of all became good friends emphasizing the tightly-knit nature of the subject, and we would like to acknowledge their contributions here.

First of all, without Tor Haugland's practical engineering genius, we (and a lot of other people) would have achieved little. Mind you his suggestion of cod's tongues as a viable breakfast during airgun trials didn't help. We are still good friends.

One name that frequently crops up in this and other published material on seismic sources is that of another old friend Anton Ziolkowski. Together we have shared numerous long evenings during trials arguing about the pros and cons of various schemes disguising our real purpose which was to improve the sales of Scottish malt whisky. As Professor of Geophysics at Delft, he has been a constant and distinguished presence in the development of marine seismic sources.

We would also like to acknowledge a number of rewarding conversations with Svein Vaage, founder of the triple club, and with Bjorn Ursin, both of whom have been in the thick of the action for a number of years.

A number of other people helped in one way or another with a comment here or an exquisite diatribe there, but we would especially like to mention Mark Loveridge and Robert Laws, who know lots about this sort of thing, and Steve Levey and the rest of the crews of the Liv and Nina Profiler for engineering ingenuity and sleeplessness above and beyond the call of duty.

We would like to thank the S.E.G. and the E.A.E.G. for permission to use some of the diagrams published in Geophysics and First Break and also Britoil plc for permission to publish some of their array trials in the first place.

Finally, we would like to thank our families for putting up with all this.

L.H.
G.P.

Woking, U.K., January 1986.

CHAPTER 1

Underlying Physics and Concepts

1.1. ACOUSTIC WAVE GENERATION TECHNIQUES

The list of available marine seismic sources is extensive. Many have stood the test of time, whilst others have fallen by the wayside, then there are the countless variations on a theme. Technical details of the sources which are currently used most are given in the appendix. This will be updated periodically to keep it in line with the times. The methods used to generate acoustic waves in marine seismic sources can be grouped into four broad categories. These primary divisions are - chemical, mechanical, pneumatic/hydraulic, and electrical.

Chemically based, or explosive sources, are perhaps the simplest in concept. They are rarely used these days, so their importance is mainly historical. In early days explosive sources were somewhat haphazard. It was almost a case of chucking a stick of dynamite over the side and hoping for the best ! The uncertainty in the location of the charge when it exploded usually necessitated the use of a second boat to tow the receiver ! These methods were superseded by a second generation of more sophisticated and carefully controlled explosive based sources. The methods of operation were quite varied. The most straightforward were solid explosive capsules with appropriately timed fuses. The dispensing of these charges was automated and relatively safe. In other solid explosive sources, the charge was detonated in an enclosed perforated cage, with the object of attenuating the secondary bubble oscillations. The problem here was that the cage suffered cumulative damage and needed frequent replacement. Then there were sources in which explosive mixtures of gases were ignited in a rubber sleeve, an exhaust system dispelling the waste gases. This latter type of source is still occasionally used, but increasingly rarely. Nonetheless an enormous volume of seismic data has been shot with explosive sources, and from time to time some of it emerges for reconsideration.

Sources which fall into the category ‘mechanical’ rely on a physical vibration, usually of a plate, to transmit seismic waves into the water. A typical example would be the marine equivalent of Vibroseis*, in which a plate is vibrated through a sweep of frequencies. Although the signature of such a source is extremely long, the bandwidth and spectrum can be controlled easily. Only recently have such sources been introduced to the marine environment, Peacock et al. (1982).

The next category is pneumatic/hydraulic. In these sources a slug of gas (usually air) or fluid (usually water) is dispelled under pressure into the water. The driving pressure can be pneumatic (e.g. compressed air) or hydraulic, and in some cases is a combination of both. The most notable example is the airgun, in which a controlled volume of high pressure air is suddenly vented into the water. The resultant rapid expansion and subsequent oscillation of the air bubble generates seismic radiation. Another example is the watergun in which a

* Registered trademark of CONOCO

slug of water is expelled under pressure. Cavitation occurs and the resultant implosion produces the radiation. Other sources in this general category rely on rather different principles, such as the expulsion of a bubble of steam. Rapid condensation of the steam produces a cavity. The resultant implosion of water generates the seismic radiation.

The final category of electrically operated sources tend to be special purpose. The principle is that a large electrical charge is stored, and then discharged through the water. Vapourisation occurs and a steam bubble is formed, this is followed by condensation and implosion. This source is known as the ‘sparker’.

The diversity of marine seismic sources discussed above produces an equal diversity in radiation characteristics. However the physical principles that govern the design and operation of all these sources are the same. Therefore in the remainder of this book these principles will be considered with as little reference to specific sources as possible.

1.2. ACOUSTIC WAVE PROPAGATION IN WATER

The only waves of concern when considering marine seismic sources are ‘elastic’ waves. That is waves which propagate with a velocity that depends upon the elastic properties of the surrounding medium. Elastic waves fall into two main categories, longitudinal or P waves in which particle motion is in the same direction as wave propagation, and S waves in which particle motion is perpendicular to the direction of wave propagation. Fluids have no resistance to shear and so will not support the propagation of S waves. Therefore only P waves need be considered.

A full discussion of the equations which describe the propagation of these waves would require a book to itself. However the important steps and physical concepts will be treated for completeness, and for those with a mathematical bent. For a more detailed discussion see for instance Batchelor (1967) or Aki and Richards (1980). In what follows the formalism of Batchelor will be followed.

For convenience, the short hand notation of equation 1.1 will be introduced :

$$\frac{D}{Dt} = \frac{\partial}{\partial t} + \underset{\sim}{u} \cdot \nabla \tag{1.1}$$

in which $\underset{\sim}{u}$ is the vector velocity of the fluid. The first fundamental equation which must be introduced is the ‘continuity’ or ‘conservation of mass’ equation. The physical principle this imbodies is simply that for each element of the fluid, the rate of change of volume must be balanced by the rate of change of density. So,

$$\frac{1}{\rho}\frac{D\rho}{Dt} + \nabla \cdot \underset{\sim}{u} = 0 \tag{1.2}$$

where ρ is the density. Equation 1.2 can be re-written as 1.3, by introducing the quantity 'a' as defined by equation 1.4 :

$$\frac{1}{\rho a^2}\frac{Dp}{Dt} + \nabla \cdot \underset{\sim}{u} = 0 \tag{1.3}$$

and,

$$a^2 = \left(\frac{\partial p}{\partial \rho}\right)_s \tag{1.4}$$

where p is pressure, and the S represents constant entropy. The second fundamental equation which must be included is the 'equation of motion', which relates the acceleration of the fluid to the forces and stresses acting upon it. The equation in its full form is called the 'Navier-Stokes equation of motion', and contains terms which account for molecular transport effects. For the fluid flow conditions under consideration it is legitimate to assume that the entropy of each element remains constant. The flow field is then said to be 'isentropic' and all molecular transport effects can be neglected. The resulting simplified equation of motion is then given by 1.5 :

$$\rho \frac{D\underset{\sim}{u}}{Dt} = \rho\underset{\sim}{F} - \nabla p \tag{1.5}$$

where $\underset{\sim}{F}$ is the vector body force experienced by the fluid.

A further simplification that can be made at this stage is to assume that the disturbance to the fluid is only slight as the wave propagates. That is, the variation in density from the equilibrium value, ρ_1 $(= \rho - \rho_0)$. the variation in pressure, p_1 $(= p - p_0)$.and the velocity $\underset{\sim}{u}$ are all small. With this simplification equations 1.3 and 1.5 can be approximated by equations 1.6 and 1.7 :

$$\frac{1}{\rho_0 a_0^2} \frac{\partial p_1}{\partial t} + \nabla \cdot \underset{\sim}{u} = \underset{\sim}{0} \tag{1.6}$$

and,

$$\rho_0 \frac{\partial \underset{\sim}{u}}{\partial t} = \rho_1 \underset{\sim}{F} - \nabla p_1 \tag{1.7}$$

in which a_0 is the value of a at $\rho = \rho_0$. The relationship of equation 1.8 may now be arrived at by differentiating equation 1.6 with respect to time, substituting the expression for ($\partial \underset{\sim}{u}/\partial t$) from equation 1.7, and using the relationship 1.4 :

$$\frac{1}{a_0^2} \frac{\partial^2 p_1}{\partial t^2} = \nabla p_1^2 - \rho_1 \nabla \cdot \underset{\sim}{F} - \frac{\underset{\sim}{F} \cdot \nabla p_1}{a_0^2} \tag{1.8}$$

Now the body force $\underset{\sim}{F}$ is due to the Earth's gravitational field, so $\underset{\sim}{F} = \underset{\sim}{g}$ and hence the $\nabla \cdot \underset{\sim}{F}$ term is zero and the final $\underset{\sim}{F}$ term of equation 1.8 is completely negligible. We are left with the spherical wave equation which is given in its more usual guise in equation 1.9 :

$$\frac{1}{c^2} \frac{\partial^2 p}{\partial t^2} = \nabla^2 p \tag{1.9}$$

The quantity a_0 turns out to be the phase velocity of wave propagation, c, in equation 1.9. The main case of interest is the point source with a spherically symmetric radiation field, so it is convenient to convert to spherical polar coordinates. If the possibility of inward travelling waves is ignored, then the solution of equation 1.9 is of the form:

$$p = \frac{1}{r} f(t - \frac{r}{c}) \tag{1.10}$$

where f is some unknown source dependent function. If K is the bulk modulus of elasticity of the fluid, then the phase velocity or speed of sound is given by equation 1.11 :

$$c^2 = \frac{K}{\rho} \tag{1.11}$$

At seismic frequencies, the velocity is essentially independent of frequency, however it is sensitive to temperature and pressure. The insensitivity to frequency (or absence of dispersion) means that the characteristic pulse shape of the radiation from a marine seismic source is maintained as the pulse propagates through the water.

Typical sound velocities in water likely to be encountered in marine seismic acquisition lie in the range 1450-1500 m/s. In general, velocity increases with temperature, with pressure, and with salinity. For example, at 3% salinity, the velocities at 5°C, 10°C, and 15°C, are respectively 1462 m/s, 1480 m/s, and 1500 m/s. In general, in shallow waters in which the temperature drops with depth, there will be a corresponding fall in the acoustic velocity. However in water deeper than about 1km the temperature is nearly constant, so the increase in pressure results in a general increase of velocity with depth.

1.3 THE RELATIONSHIP BETWEEN VELOCITY, PRESSURE AND ENERGY

The relationship between velocity and pressure may be derived as follows. Differentiate the expression for pressure of equation 1.10 with respect to r and substitute the result into equation 1.7. If it assumed that the body force $\underset{\sim}{F}$ is zero and that the distance from the source is large enough for $\tilde{1}/r^2$ terms to be neglected, then the expression of equation 1.12 results :

$$u = \frac{1}{\rho c}\, p \tag{1.12}$$

The kinetic energy of a particle of mass m, moving at a speed u, is simply $\frac{1}{2}mu^2$. The potential energy in a sound wave is equal to the kinetic energy as shown for instance by Coulson (1941). E , the total rate of energy flow through a shell at a radius r from the source can now be calculated as the area of the shell times the speed of flow times the total energy density (potential plus kinetic). E is therefore given by equation 1.13 :

$$E = 4\pi r^2\, c\, \rho u^2 \tag{1.13}$$

An expression for the total energy emission from the source, T, can now be derived by combining equations 1.12 and 1.13, recognising that the pressure p is in fact a function of time, p(t), and integrating over time.

$$T = \frac{4\pi r^2}{\rho c} \int_0^\infty p(t)^2 \, dt \qquad (1.14)$$

Having derived the relationship between pressure and energy, it is worth considering how these parameters vary with distance from the source. If it is assumed that there is no absorption of energy as it propagates through the water, then the law of conservation of energy states that the total rate of energy transfer through any sphere of radius r from the source, must be constant. The area of a sphere is simply $4\pi r^2$, so the energy transfer rate per unit area must decrease as the inverse of the square of the distance from the source. At large distance from the source (the far field), where equation 1.14 holds, the pressure of the wave will decrease as the inverse of the distance from the source.

1.4. BUBBLE MOTION

A substantial number of seismic sources produce their acoustic radiation by the release of high pressure gases into the water. This may be directly, as in airguns, for example, or indirectly, as in explosive sources, in which gases are produced as a result of chemical reactions. The generation of acoustic energy is closely coupled to the high speed motion of these gas bubbles.

The behaviour of a bubble of high pressure gas when it is released suddenly into water is illustrated in the top half of Figure 1.1, in which bubble radius is plotted against time. Initially the pressure inside the bubble greatly exceeds the external or hydrostatic pressure, so the gas bubble expands rapidly. The momentum of the expanding bubble is sufficient to carry the expansion well beyond the point (when the bubble radius = R_0) at which the internal and hydrostatic pressures are equal. When the expansion ceases, the internal bubble pressure is below the hydrostatic pressure, so the bubble starts to collapse. The collapse overshoots the equilibrium position and the cycle starts once again. The bubble continues to oscillate, with a period typically in the range of tens to hundreds of milliseconds (the corresponding radii being in the range 0.1 - 1 metre). The oscillation would continue indefinitely if it were not for frictional losses, and the buoyancy of the bubble which eventually causes it to break the surface. The cyclic compression and rarefaction in the water surrounding the bubble as it oscillates generates acoustic radiation. The form of this radiation as a pressure wave is illustrated in the lower half of Figure 1.1, and is essentially a periodic damped oscillation.

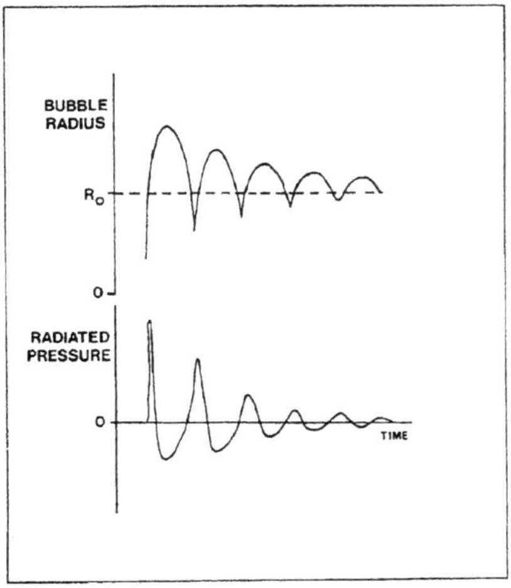

Figure 1.1 :
The top graph shows the radius of an oscillating bubble as a function of time. The lower graph is the corresponding radiated pressure. R_0 is the radius at which the bubble's internal pressure is equal to hydrostatic pressure.

To a first approximation the period of oscillation of the bubble is described by equation 1.15 :

$$T_{osc} = C1 \frac{P^{1/3} V^{1/3}}{P_0^{5/6}} \tag{1.15}$$

in which P and V are the initial internal pressure and volume of the gas before it is released, and P_0 is the hydrostatic pressure. The constant C1 depends upon the details of the source design. Detailed experimental and theoretical work on airguns (see Vaage et al. (1983) and references therein) has produced the following expressions for the primary amplitude of the emitted pulse :

$$A = C2\ V^{1/3} \tag{1.16}$$

when the pressure and depth are held constant, and

$$A = C3\ P^{3/4} \tag{1.17}$$

when the volume and depth are held constant. Once again the constants of proportionality C2 and C3 are functions of the source design.

The detailed physics of seismic source bubbles is complex, and beyond the scope of this work (see Ziolkowski and Metselaar (1984) and references therein). The parameters of the bubble as a function of time are really quite surprising. The temperature can drop to a hundred or so degrees Celsius below zero, causing ice crystals to form in the bubble. There is also evidence that the bubble is constituted of a foam of small bubbles rather than being one large one.

1.5. THE GHOST OR VIRTUAL IMAGE

The time domain pressure wavelet which characterises the radiation emitted by a seismic source is referred to as the source signature. This is a misnomer in that there is no single wavelet which fully describes the source radiation (as will be seen later). However the term 'the source signature' is commonly used, without specifying position in the wavefield, to describe the vertically travelling pressure wavelet. To avoid confusion, it is as well to be more specific, and always call this special case the vertically travelling far field signature.

The reflection of the source radiation at the sea surface has an inextricable effect on the radiation field, and as such is considered to be an intrinsic feature of the source wavefield, from the point of view of data processing. Source signatures, therefore, are usually presented with this intrinsic reflection component included.

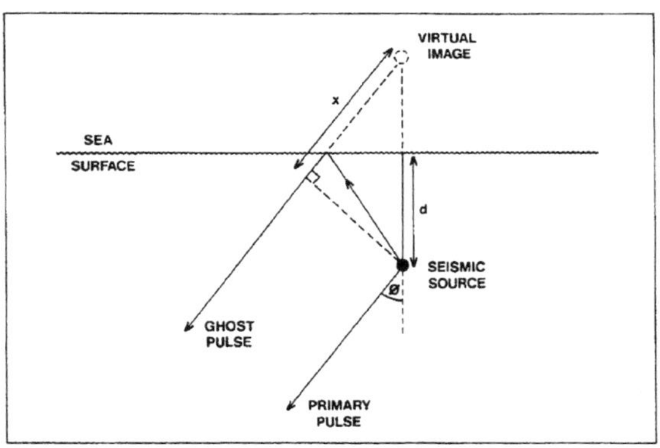

Figure 1.2 :
The marine ghost pulse, which results from the sea surface reflection, appears to originate from the virtual image of the seismic source. It is delayed in time by x/c with respect to the primary pulse, where c is the speed of sound in water.

At seismic frequencies, a planar water/air interface is generally an almost perfect reflector. In some circumstances this is not the case as will be seen in Chapter 4. However for the discussion of this section a sea surface reflection coefficient of -1 will be assumed throughout. The minus sign represents a polarity reversal of the reflected waves. In essence, therefore, seismic energy generated beneath the surface of the sea cannot escape into the air. Acting like a mirror, the sea surface reflection produces a virtual image of the seismic radiation which is generally called the 'marine ghost'.

Considering Figure 1.2, the ghost pulse or signature will be delayed in time with respect to the primary pulse by x/c, where c is the velocity of sound in water. The time delay, τ , is therefore given by equation 1.18 :

$$\tau = \frac{2d\cos(\phi)}{c} \tag{1.18}$$

where d is the source depth. The time delay is a function of the angle, ϕ. In fact, the majority of the seismic energy recorded in seismic experiments will have originated at angles very close to the vertical, giving this direction special significance. In the vertical direction the ghost pulse will be delayed in time by 2d/c, where d is the depth of the source. The composite wavelet of energy is therefore the summation of a primary wavelet, and a ghost wavelet delayed in time and reversed in polarity with respect to the primary, as shown in Figure 1.3.

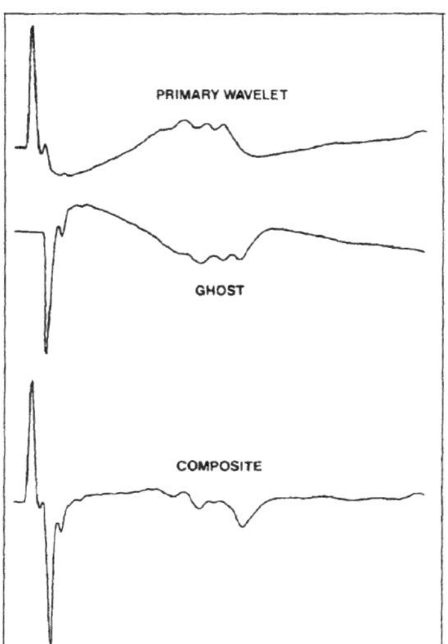

Figure 1.3 :
The ghost wavelet is delayed in time, and has opposite polarity to the primary wavelet. The composite source signature is the summation of primary wavelet and ghost.

Figure 1.3 demonstrates the effect of the ghost on the time domain wavelet. The consequences of this in the frequency domain should also be examined.

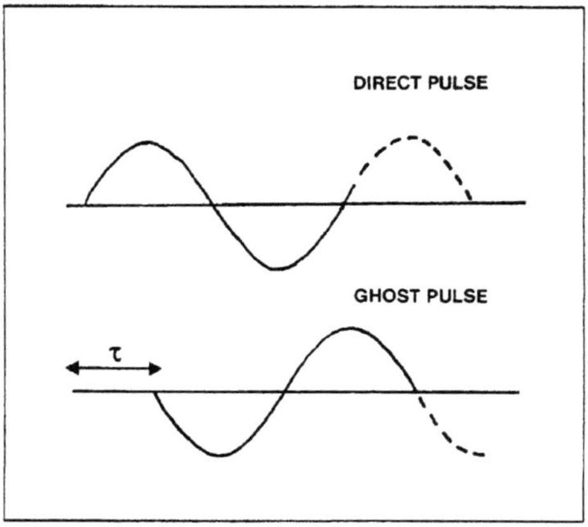

Figure 1.4 :
The ghost pulse is reversed in polarity and has a time delay τ with respect to the direct pulse. The exact value of τ determines whether the interference between the pulses is constructive or destructive.

Consider Figure 1.4, in which the seismic source signature is a sinusoid. The ghost pulse is reversed in polarity, and delayed in time by τ , where τ is given by equation 1.18. The distance travelled in time τ will be $\tau c = 2d\cos(\phi)$. Clearly, if $2d\cos(\phi)$ is half a wavelength, then the direct and ghost waves will add constructively, whereas if $2d\cos(\phi)$ is a full wavelength then the two sinusoids will add destructively. So in summary :

Destruction if : $2d\cos(\phi) = 0.0\lambda,\ 1.0\lambda,\ 2.0\lambda, \ldots$
Construction if : $2d\cos(\phi) = 0.5\lambda.\ 1.5\lambda,\ 2.5\lambda, \ldots$

However, of course, the typical seismic source signature will not be sinusoidal, but will contain components at a range of frequencies. The minimum and maximum interference conditions above will therefore modulate the complete continuous amplitude spectrum with the function shown in Figure 1.5.The frequencies, ν_n , of the ghost notches are given by equation 1.19 :

$$\nu_n = \frac{nc}{2d\cos(\phi)} \tag{1.19}$$

in which n is a positive integer, c is the speed of sound in water and d is the source depth.

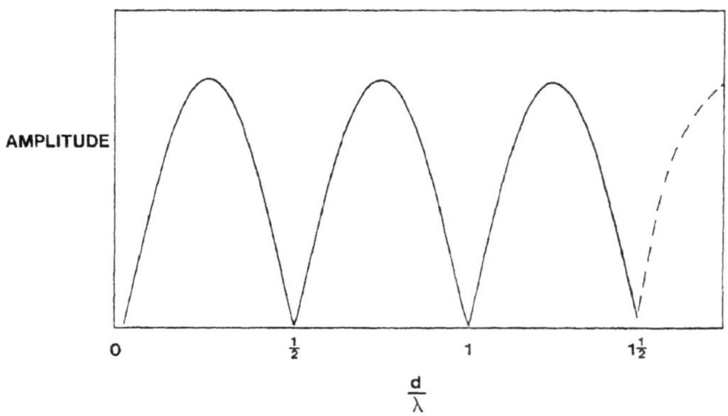

Figure 1.5 : Amplitude template introduced by the source ghost. Amplitude is plotted as a function of d/λ , where d is the source depth and λ the radiation wavelength.

The effect of the ghost on the frequency content of the seismic signal is therefore substantial. Figure 1.6 shows the vertically travelling far field signatures for the same source operated at a variety of depths. The signatures are shown on the left of the figure and the corresponding amplitude spectra (on a linear scale) are shown on the right. The correct relative amplitudes have been preserved throughout. The increased ghost delay is noticeable in the signatures as the depth increases, however the effects on the amplitude spectra appear much more dramatic. For instance, at 12m the ghost notches heavily modulate the spectrum, compared to 3m, say. However the low frequency amplitude (0-10Hz) at 3m is a factor of 2 or 3 below that at 12m. Choice of depth is therefore a most important aspect of source design, and will depend on the type of source being used and the overall aims of the survey. For instance, if deep penetration is required, the low frequency content of the signature should not be compromised (high frequencies are quickly attenuated in the Earth, so penetration to depth can only be achieved with low frequencies), and a deep source should be used. Conversely, if high resolution in the shallow is required, then a shallow source will generally be used to extend the high frequency bandwidth. Other effects compete with this ghost effect. For instance, from equation 1.15 it is clear that the greater hydrostatic pressure at greater depth reduces the bubble oscillation period. This reduces low frequencies with increasing depth in direct opposition to the ghost effect. This secondary effect is however completely dominated by the ghost effect.

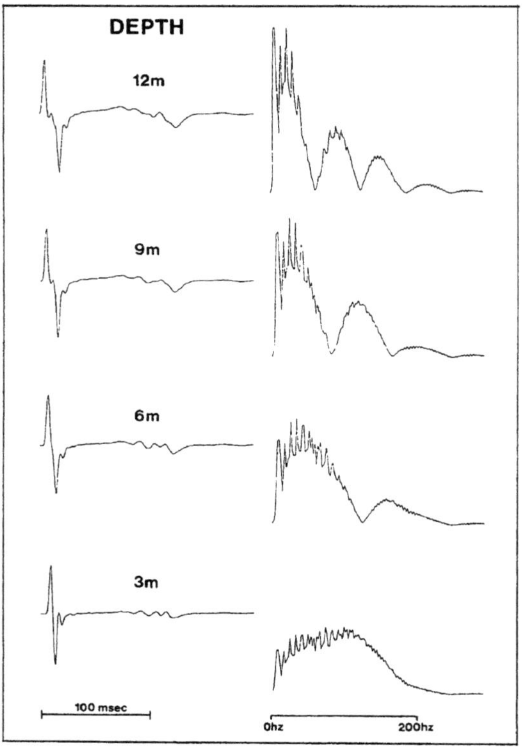

Figure 1.6 :
Variation of the vertically travelling far field signature and its spectrum as a function of source depth.

1.6. THE NEAR AND FAR RADIATION FIELDS

Unfortunately there is some confusion in the use of the terms 'near field' and 'far field'. They do have quite precise definitions, however it is common to find them used in a much looser sense. The precise definitions relate to effects which are only important close to the source. For the purposes of this discussion it will be convenient to rewrite equation 1.10 as follows :

$$p = \frac{1}{r} f' \, (t - \frac{r}{c}) \tag{1.20}$$

where f' is the time differential of some function f, p is pressure, r is radial distance from the source and c is the speed of sound in water. Now Newton's second law may be written as in equation 1.21 :

$$a = - \frac{1}{\rho} \frac{\partial p}{\partial r} \tag{1.21}$$

in which a is acceleration, and ρ is density. Differentiating equation 1.20 and substituting the result in equation 1.21 gives :

$$a = \frac{1}{\rho r^2} f'(t - \frac{r}{c}) + \frac{1}{\rho c r} f''(t - \frac{r}{c}) \tag{1.22}$$

Finally the expression for the particle velocity is obtained by integrating equation 1.22 with respect to time, to give :

$$u = \frac{1}{\rho r^2} f(t - \frac{r}{c}) + \frac{1}{\rho c r} f'(t - \frac{r}{c}) \tag{1.23}$$

Considering the above expression, the 'near field' may be defined as that region in which the $1/r^2$ term dominates - that is close to the source. The 'far field' is then the region distant from the source in which the 1/r term dominates. It is interesting to note that the <u>shape</u> of the particle velocity wave is dependent on distance, because it is the sum of two terms that depend differently on distance. On the other hand there is only a single distance dependent term in the expression for pressure (equation 1.20), so the shape of the pressure wave is independent of distance.

A common rather different use of the term far field considers it to start at that point at which the pressure signature is indistinguishable (i.e. within some error tolerance) from the signature that would be measured at infinity in the same direction. The exact position adopted depends on many factors, not least opinion ! However more scientifically, it depends upon the acceptable error tolerance, source depth, frequency of interest, direction and source spatial extent. For example, many would consider 100m to be in the far field of a point source deployed at 5m depth. However at 100m the amplitudes of ghost and direct arrivals will differ by 10 percent, whereas in the true far field they are identical ! Whatever error criterion is used, the far field of spatially extended sources starts much further away than that of point sources.

The term 'far field signature' is often used with the implicit understanding that it refers to the vertically travelling signature. This is rather a misnomer. In its broader use far field signatures exist for all directions. In addition note that there is no such thing as the far field signature since all sources are directional, even so-called 'point sources' by virtue of the ghost.

1.7. THE MEASUREMENT OF SIGNATURES

A detailed knowledge of the radiation fields of seismic sources is essential to deciding which source to use for a particular purpose, and how to deploy it. The same detailed knowledge is required to enable the wavefield to be deconvolved when processing the data. The source may be directional and/or interacting, and this may need to be taken into account in the processing. However, to a first approximation at least, the source wavefield can be characterized by the vertically travelling far field signature. An accurate knowledge of this signature is therefore mandatory. Estimates of the vertically travelling far field signature may be obtained by hydrophone measurements in the far field, or by extrapolating near field measurements to the far field.

Far field measurements are notoriously difficult and expensive. They must be carried out in deep water to ensure that the hydrophone is indeed in the far field. The biggest problem is aligning the hydrophone vertically beneath the source. A hydrophone cable of perhaps 150m long will not hang vertically in the water, especially if there are cross currents, or if the boat towing the hydrophone is moving. An alternative to towing the hydrophone is to attach it to a stationary buoy, and steam past it towing the source. Again it is impossible to tow the source to a point vertically above the hydrophone especially if the source is extended. Triangulation techniques can be used to locate the hydrophone. For example, by using travel times from extreme sources on an extended array, or by using acoustic 'pingers' outside the seismic band since hydrophones have a much broader bandwidth than is required. So with care, reasonable estimates of the vertically travelling far field signature can be obtained.

Near field measurements also have their problems. The hydrophone must be robust to survive so close to the source. Its sensitivity must also be chosen carefully to avoid it being over- or under-driven. However suitable equipment does exist and good near field measurements are made routinely. The signature at any position in the wavefield can be deduced from near field measurements as will be seen.

The law of superposition of wavefields states that two or more waves can traverse the same space independently of each other, and that the particle displacement at any point is simply the vector addition of the displacements caused by the independent waves. So, assuming the principle of superposition, the signature at any position in the wavefield of a marine seismic source can be calculated as follows. The direct wavelet from each element of the source array is calculated for the required position by delaying the near field signature and including amplitude scaling according to the 1/r law. Similarly all the secondary ghost wavelets are computed. The composite signature is simply the direct sum of all these components. For marine seismic sources the superposition principle has been shown to work well for single sources and for non-interacting arrays of sources. The wavefield of an interacting array cannot be obtained by superposing the near field signatures of the individual array elements fired in isolation. However, as will be shown in Chapter 3, it is possible to derive a set of equivalent or ‘notional’ sources which can be used with the law of superposition to give the correct result.

1.8. RELATIVE MOTION EFFECTS

If the signature of a single element seismic source is known for some fixed distance, then the signature to be expected at some other fixed distance can be computed using the 1/r amplitude scaling law. For example, the calculation of far field signatures from near field measurements, as discussed in the preceding section. However there are often relative motion effects between seismic source and hydrophone which must be allowed for. This relative motion arises because the hydrophone moves through the water at the speed of the towing vessel, which is typically 2-3 m/s. However for many seismic sources, the source of the radiation is released into the water and thereafter follows its own peculiar motion. For example, in the case of an airgun, a bubble of air is released into the water with an initial forward velocity which is the same as that of the towing vessel. However the drag on the bubble causes this forward motion to slow rapidly. Furthermore the buoyancy of the bubble causes it to rise in the water. This upward motion is complex and non-linear, in that the rapid expansion and contraction of the bubble modulates the drag and hence the rise velocity. Without such motions, the rise velocity should be constant in the manner of a spherical cap bubble, (Batchelor (1967)).

For a far field hydrophone, the relative motion between source and receiver is usually negligibly small. However in the near field the effects can be large. As an illustration consider Figure 1.7. In this example it is assumed that the seismic source is a gas bubble which has upward motion at constant velocity, but no forward motion. On the other hand, the near field hydrophone has a constant forward velocity. Plotted in the figure is $r(t)/r(t_0)$, where $r(t)$ is the functional variation of separation with time, and $r(t_0)$ is the separation at time zero. In case (a) the velocities are zero and there is no functional variation. In cases (b) and (c) however, the variation is large. For example, in case (b), at about 300 msec, the separation is less than half that at time zero. The measured signal amplitude will therefore be more than double the value that would be measured with no relative motion.

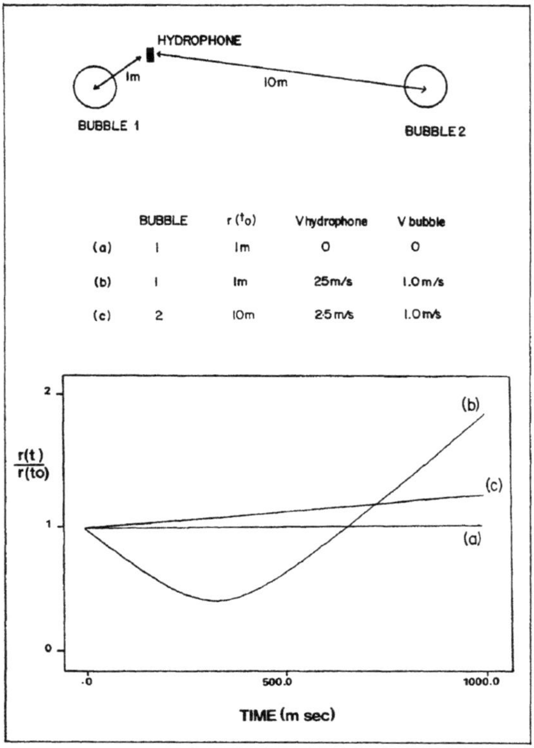

Figure 1.7 : Relative motion effects between hydrophone and seismic source (bubble) for three cases as shown. The hydrophone motion is to the left and the bubble motion upwards. $r(t_0)$ is the hydrophone to bubble separation at time zero, and $r(t)$ is the functional variation of separation with time.

To be of value, near field measurements must be referred to some fixed distance, so the measured amplitudes must be corrected for the separation effect illustrated in Figure 1.7. For seismic sources which produce gas bubbles, a linear relative velocity model is an excellent first approximation. However if even greater accuracy is required, then acceleration terms must be included. Such accuracy has not been found necessary in practice.

Figure 1.8 (a) is a measured near field signature of an airgun source. The hydrophone was placed 1m from the source (above and behind it). The relative motion effects are visible particularly in the second and third negative excursions of the signature, for which the amplitude is too large because of the reduced source/receiver separation at this time. Figure 1.8 (b) shows the same signature corrected for relative motion effects. The damped decaying oscillation is now as expected.

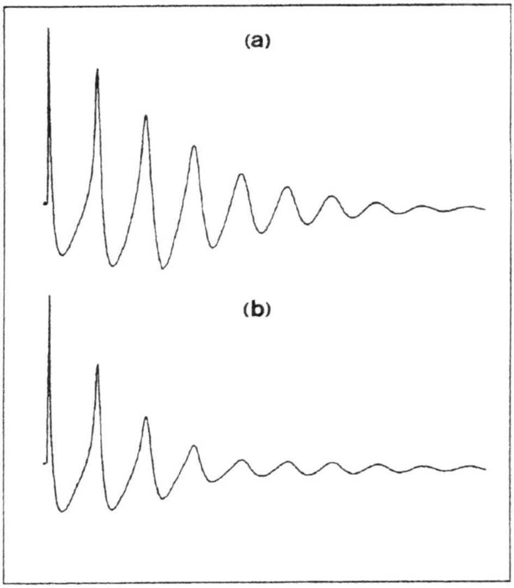

Figure 1.8 :
(a) The measured near field signature of an airgun.
(b) The same signature corrected for relative motion effects (hydrophone forward velocity of 1.8 m/s and bubble rise velocity of 1.0 m/s).

1.9. SIGNATURE PARAMETERS

There are a number of useful measures commonly used to describe far field signatures. The first is 'peak to peak amplitude'. This is simply the amplitude difference between the peak of the primary event and the peak of its ghost - the amplitude A1 shown in Figure 1.9. Depending on the type of source, the primary peak may not be at the start of the signature. When specifying peak to peak amplitude, the bandpass filter which has been applied to the signature must also be specified, since there will clearly be a functional dependence on filter setting.

Another descriptive parameter is 'primary to bubble ratio'. The bubble part of the signature is composed of the secondary oscillations that occur after the primary event. The secondary peak to peak amplitude is A2, as shown in Figure 1.9. The primary to bubble ratio is then A1/A2. Again the filter setting must be specified.

In geophysics, the traditional unit for pressure is the bar (1 bar = 10^5 N/m^2 = 14.5 psi, or approximately 1 atmosphere). The amplitude (or excess pressure) of a signature is normally measured in bar-metres (b-m). Distance is usefully parameterised in these units. An amplitude of A bar-metres means that at a distance of 1 metre from the source the amplitude value would be A bars. So if the pressure measured at a distance r from a source is P(r), then using the 1/r scaling law, the corresponding pressure in bar-metres will be P(r).r.

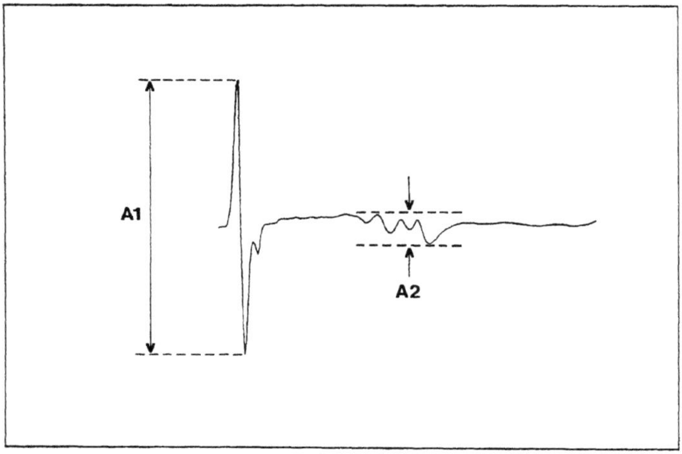

Figure 1.9 : Commonly used signature parameters -
(1) The peak to peak amplitude, A1.
(2) The primary to bubble ratio, A1/A2.

1.10. FACTORS DETERMINING CHOICE OF SOURCE

A wide range of factors may influence choices of source for specific applications. Not least of these are practical considerations such as which sources are available at the required time and place in the world, and how expensive they are to use. In some geographical areas the choice may be limited by operational factors. For instance, in shallow water larger boats and source systems may be impossible to use. There may also be environmental considerations, although modern sources are relatively 'pollution' free. There were more problems in the past with explosive sources, for instance, which were sometimes banned because of their destructive effect on the local fish population ! Aside from such practical considerations there are various design specifications that the source must meet to fulfill the scientific objectives of the survey in question.

In the time domain, the traditional criteria for assessing the pulse are the pulse duration, the primary to bubble ratio, and the pulse repeatability. It can be argued that the ideal profile is an impulse response, so in practical terms short duration and high primary to bubble ratio are considered desirable. Such an approach has the advantage that the need to deconvolve a complex pulse shape diminishes. Furthermore the source signature is often not known accurately, and may even vary with time. Repeatability and stability of the pulse is therefore another prime consideration. In source systems in which the source is stable and the radiation field is well determined, the signature deconvolution becomes much less of a problem and these 'classical' time domain constraints can and should be relaxed.

More modern thinking, coincident with much improved techniques for predicting the far field radiation pattern, would place frequency domain characteristics and energy output as the primary considerations. Since higher frequencies are rapidly attenuated in the Earth, high energy, low frequency sources are required if the objective is penetration to depth. Conversely less powerful high frequency sources are used when the objective is high resolution in shallow data. If it can be predicted, the time domain response of the source is irrelevant as the convolutional model (see later) is an excellent basis for source signature correction. Note finally that it is even sometimes possible to tune the source amplitude spectrum in detail so that the resultant spectrum at the target (after ghosting and attenuation in the Earth, etc.) is as desired.

CHAPTER 2

Source Arrays and Directivity

2.1. THE REASONS FOR USING ARRAYS

It is common for marine seismic sources to be deployed in groups or arrays. This is done for a variety of reasons. The most obvious of these is simply to increase the power of the source. The alternative of using increasingly larger sources is impractical because size increases of a single source are inevitably accompanied by changes in the characteristics (e.g. bandwidth) of the emitted pulse. In essence, therefore, a source system of 'n' times the power of a single source can be achieved by firing 'n' sources together as an array, whilst maintaining the basic pulse characteristics of the single source.

The next progressive step is to use a variety of different sources in the array. The possibilities here are endless. Low and high frequency sources may be used together to improve the bandwidth. Sources which have deep notches in their spectra may be combined with others that have complementary peaks, to produce an overall flatter spectrum. Array elements may be placed at different depths in the water to avoid large ghost notches in the spectrum. There is also the possibility of firing the array elements at different times to maximize energy emission in directions other than the vertical (beam-steering techniques will be treated later).

Unfortunately, life is never simple and a variety of other array phenomena must be considered. Of these, the most important are interference effects between the radiation fields of the individual array elements, and interaction effects between the physical sources of the radiation (this will be considered in the next chapter). Both these phenomena, whilst introducing complications, can be of benefit.

2.2. PULSE SHAPING

One of the commonest applications of arrays is to control the shape of the time domain pulse (see for example Nooteboom (1978)). The typical marine source signature is composed of a primary pulse which is preceded and/or followed by various other oscillations, with periods and timescales dependent upon the type and size of the source. These oscillations are often considered undesirable in that they detract from the ideal situation of a short sharp impulse. By combining sources with different secondary pulse characteristics in an array, these unwanted oscillations can be attenuated. The composite pulse will then have a more desirable shape (in the classical sense) than the individual elements. Such arrays are called 'tuned arrays'. The principle is illustrated in Figure 2.1, in which several airguns are combined such that the primary pulses add constructively, whereas the periodic secondary oscillations add up destructively.

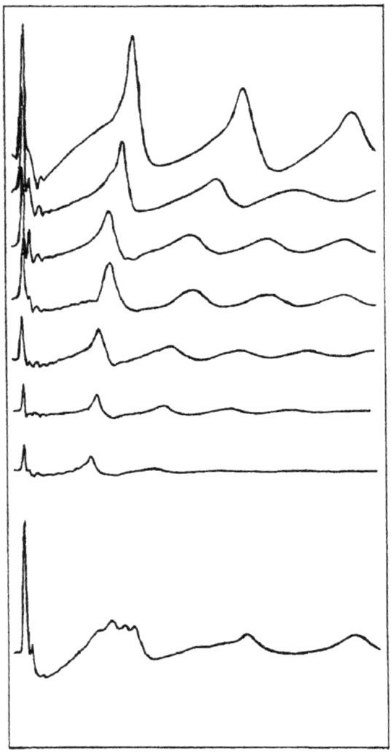

Figure 2.1 :
The bottom signature is the summation of the top seven signatures. Notice how the primary event adds constructively whereas the secondary oscillations add destructively. (Note : the true amplitude of the composite signature is three times that shown).

It is certainly true that a source signature closely resembling an impulse response reduces the need of complex signature deconvolution. This powerful argument has quite rightly gained almost universal acceptance. However pulse shaping in an array does destroy potentially useful energy, and it can be argued that provided the signature is known accurately, it can be successfully deconvolved, no matter what it looks like.

2.3. THE PRINCIPLES OF DIRECTIVITY

By definition the radiation emitted by a point source in a homogeneous medium has spherical symmetry. In such a medium, for example water, the near field radiation of a single element seismic source is indeed close to the spherically symmetric. However in the far field, about half the energy in the propagating pulse originates from the sea surface reflection, or ghost. Now the time delay between the primary and ghost pulse is a function of angle. This is purely a geometrical effect. So in the far field, even the simplest of marine seismic sources, the point source, has a wavefield which is directional. More commonly this directionality effect is called directivity.

Consider Figure 2.2, in which the radiation pulse emitted at an angle ϕ from the vertical, is a combination of the direct pulse and the secondary pulse caused by the sea surface reflection. The time delay between the two pulses, τ, will be the time taken to travel the distance x in water, and is given by equation 1.9. Since τ is a function of ϕ, the emitted pulse will be a function of direction. The minimum delay will be zero when $\phi = 90°$ and the maximum delay, corresponding to the travel time through twice the depth of the source, occurs when $\phi = 0°$.

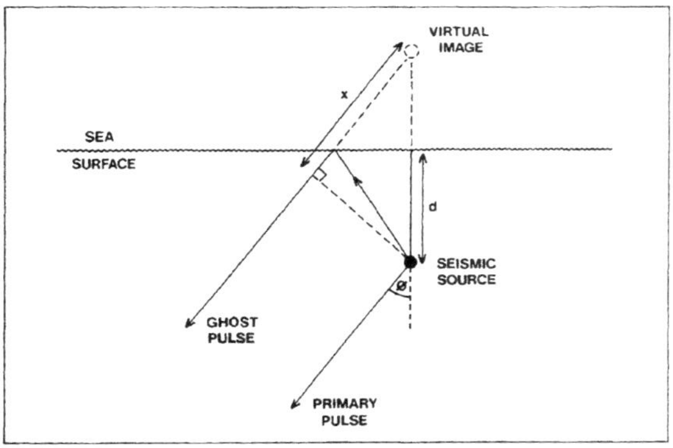

Figure 2.2 : The ghost pulse from a single element seismic source is delayed in time with respect to the primary pulse by x/c, where c is the speed of sound in water. This delay depends on emission angle ϕ, so this simplest of sources is directive.

The spatial separation of two sources produces directivity in precisely the same way as the separation of source and virtual image. However the directivity of arrays can be much more extreme. The variation of emitted signature with direction is simply a manifestation of wavefield interference effects. Most readers will be familiar with the fringe patterns produced by mono-chromatic light sources. The directivity of seismic sources is precisely the same phenomenon. However, in general, source configurations are 3-dimensional, and their emission spectra are continuous, so the resultant fringe patterns can be extremely complex. As a rule of thumb, these interference effects will be strong at radiation wavelengths that are shorter than about the total spatial extent of the source. For example, if the array size is 100m, then interference effects will be visible at wavelengths less than 100m, which corresponds to frequencies greater than 15 Hz in water.

Consider two sources separated by a distance, s, as shown in Figure 2.3. In some direction ϕ, there will be a phase delay, dt, between the primary pulses of the two sources, where dt corresponds to the travel time across the distance x; dt is given by equation 2.1 :

$$dt = \frac{s \sin(\phi)}{c} \tag{2.1}$$

in which c is the speed of sound in water. Again the time delay is a function of ϕ, and will be zero when $\phi = 0°$, and a maximum of s/c (which equals the travel time across the separation s) when $\phi = 90°$. Clearly from equation 2.1, the potential maximum phase delay is directly proportional to the source separation. Therefore at $\phi = 0°$ the pulses add constructively, whereas at other angles destructive interference takes place. In general, the greater the geometrical spread of the array, the greater will be its directivity.

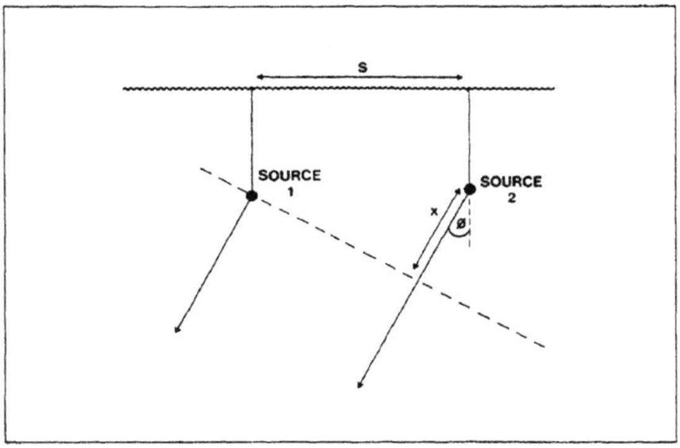

Figure 2.3 :
In arrays, the time delay between the pulses from neighbouring sources is x/c. This delay is a function of angle ϕ , so the wavefield of the array is directional.

It is clear from Figure 2.3 that when x is an exact number of wavelengths, the energy at the corresponding frequency will add constructively, so there will be a maximum in the energy emission. Similarly, when x corresponds to an exact half number of wavelengths (e.g. 1.5, 2.5, 3.5, etc) then there will be minima in the energy emission. So, at any particular frequency, there will be a mainlobe of emission in the vertical direction, and a family of sidelobes of emission at angles as described above. The direction of any particular order of sidelobe will be a function of frequency.

Perhaps the most important law that governs directivity is the trusty old law of conservation of energy. Assuming interaction effects do not come into play, then the total energy emission by a source array will be independent of array configuration. The configuration determines how this energy is distributed with angle. So if a particular configuration attenuates energy in one direction, then the 'lost' energy will inevitably appear in some other direction. The energy cannot be destroyed, just redistributed. Furthermore the pressure wavelet travelling in the vertical direction (assuming the sources are synchronized in time) will also be independent of array configuration. This follows directly from the law of superposition of wavefields. However when the array elements are close enough to interact, the situation is more complicated, and the vertically travelling pressure wavelet can be a sensitive function of array geometry. In fact, in the limiting case, the energy in the vertically travelling wavefield of an interacting array can be a factor of two less than the vertical energy from the same sources fired in a non-interacting configuration. These effects will be discussed in more detail in the next chapter.

2.4. DIRECTIVITY DISPLAYS

Directivity functions are a graphical representation of the energy emission characteristics of an array. In general, the wavefield of an array will be a continuously varying function of spatial coordinate (x,y,z). The wavefield itself can be represented in terms of amplitude against time (i.e. the source signature), or perhaps more familiar to the geophysicist, in terms of signal amplitude and phase against frequency. In essence, therefore, full directivity functions are 6-dimensional, and extremely difficult to visualise.

Meaningful directivity displays can be produced by fixing the values of several of the variables and displaying the resultant functional variation of the others. For example, if space is considered in terms of polar coordinates (r, θ , ϕ), then r may be fixed at infinity (i.e. the far field case), and the variation of the emitted signature plotted as a function of ϕ , for fixed θ . Alternatively ϕ may be fixed, or else signal amplitude may be plotted instead of the signature itself.

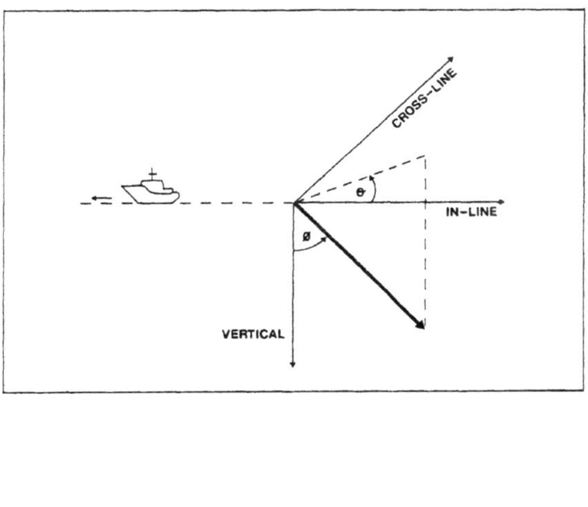

Figure 2.4 : Definitions of the angles θ and ϕ - the azimuth and angle of dip.

Before continuing a suitable system of spherical polar coordinates must be defined. In Figure 2.4 the direction of a vector in space is defined by the angles (θ , ϕ). The angle of dip, or polar angle, ϕ, is the angle between the vector and the vertical. The azimuth, θ , is the angle between the projection of the vector on the sea surface and the in-line direction.

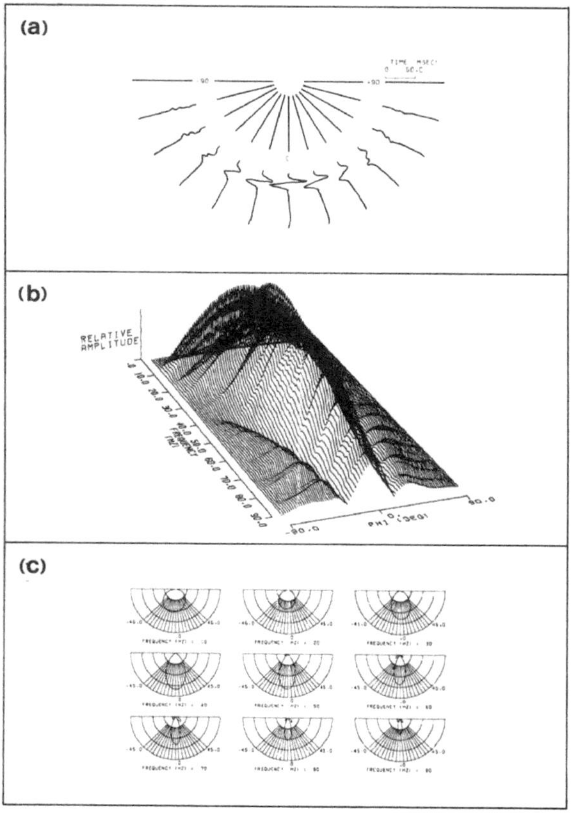

Figure 2.5 :
Alternative directivity function plots.
(a) The emitted signature versus direction.
(b) The signal amplitude versus frequency and direction.
(c) The signal amplitude versus direction at fixed frequencies.

Figure 2.5 shows three alternative ways of displaying directivity information for the same array (in this case a 40m X 40m wide array). All these plots consider the far field case of r = infinity, fix the azimuth at $\theta = 0°$, and then consider the variation of the emitted energy with angle of dip, ϕ . In other words these functions show the emitted energy distribution for the vertical plane that lies beneath the line along which the boat is steaming (i.e. the in-line plane). Plot (a) is of emitted signature versus angle of dip, and shows clearly how the amplitude and phase of the signature varies with direction (and after all it is these signatures that are convolved into our seismic sections). Plot (b) is a 3-D representation in which the z-axis is signal amplitude plotted against frequency and angle of dip. This plot gives a feel for the overall emitted energy distribution. Plot (c) is the well known polar directivity function in which a further variable, the frequency, is fixed, in this case at 60 Hz. This polar plot is essentially a slice through the 3-D plot above it at fixed frequency.

The 3-D plot of Figure 2.5 (b) is instructive in that it illustrates some general characteristics of array directivity :

(1) The main beam of emission is vertically downwards (unless the array is beam-steered - see next section).
(2) Directivity is a function of frequency. There is more at higher frequencies.
(3) Sidelobes are evident at higher frequencies.
(4) Sidelobe direction is a function of frequency, so although a sidelobe can be large at one particular frequency, over a range of frequencies the effects smooth out. This is not true of the mainlobe which is coherent.

2.5. BEAM STEERING

The elements of an array are usually fired simultaneously, producing an energy field which has a maximum in the vertically downwards direction. However in some (rare) circumstances it may be beneficial to adjust the timings of the array elements to maximise the energy emission in some other direction. This technique is called beam steering. For instance, if the predominant dip of the survey is β degrees, and there is no significant heterogeneity, then the maximum illumination of the sub-surface reflectors will be obtained if the source energy emission is maximised at β degrees from the vertical.

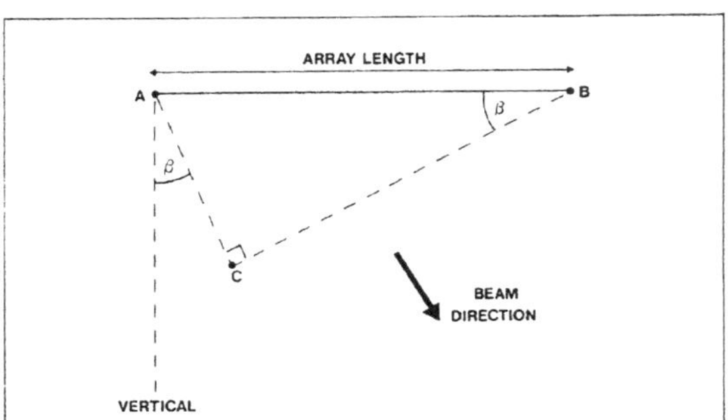

Figure 2.6 : Beam steering at an angle β may be achieved by introducing time delays along AB such that the wavefront aligns along CB. For example, the element at B is fired as the pulse from A reaches C.

Consider Figure 2.6, in which AB represents the extent of the array, and β is the required beam steering angle. The array element at position A is fired first, and the element at B is fired last. The time delay, dt, corresponds to the travel time in water between A and C. In other words, B is fired as the pulse from A reaches position C. This alignment of pulses along CB will maximise the energy emission (maximum constructive interference) at angle β from the vertical. The required time delay in seconds/metre is given by equation 2.2 for the general case:

$$\frac{dt}{dx} = \frac{\sin(\beta)}{c} \tag{2.2}$$

where c is the speed of sound in water. Figure 2.7 shows directivity functions for an array which has been beam-steered at 20°, corresponding to dt/dx = 0.23 msec/m.

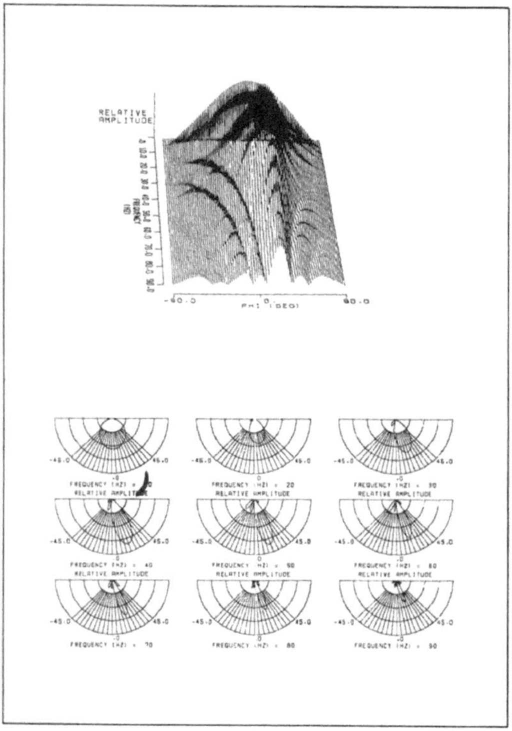

Figure 2.7 :
Directivity functions for an array which has been beam steered at 20° . This corresponds to a linear time delay along the array of 0.23 msec/m.

2.6. DIRECTIVITY OF LONG ARRAYS

A long array is one in which the spatial extension is primarily in-line, which corresponds to azimuth, θ = 0°. An example of a long array and its characteristic directivity functions is shown in Figure 2.8. This configuration consists of six sub-arrays of sources, each separated by 30m, to give a total array length of 170m. Emitted signature directivity functions are shown for the in-line (θ = 0°) and cross-line (θ = 90°) planes. The in-line plane has special significance in that it contains the source and receiver arrays. The cross-line plane shows the other extreme of directivity, however it should be stressed that the directivity will vary continuously at intermediate planes between these two extremes.

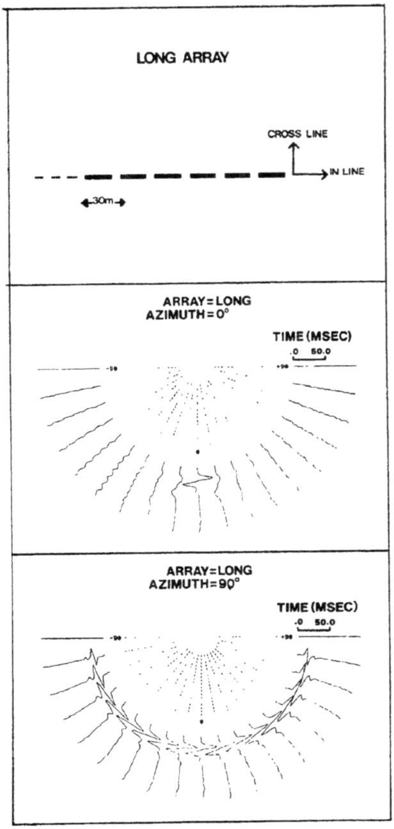

Figure 2.8 :
Directivity functions for a typical long array. Azimuth = 0° corresponds to the in-line plane, azimuth = 90° to the cross-line plane.

The directivity functions are as expected. In the in-line plane, where there is large geometrical dispersion, the emitted signature varies dramatically with direction. In the cross-line direction the array looks like a point source, so there is only a small variation caused by the ghost. So a long array has a narrow vertical beam of emission in the in-line plane, which gradually increases in size as the azimuth changes from 0° to 90°. Cross-line the beam is very wide with large amounts of energy being emitted at high angles of dip. Figure 2.9 further emphasises the directivity in the in-line plane. The first 50msec of the signatures are shown at high angular resolution around the vertical direction. The signatures are plotted at 1° intervals out to ± 15°. Clearly, the amplitude and phase of the signature is very sensitive to angle.

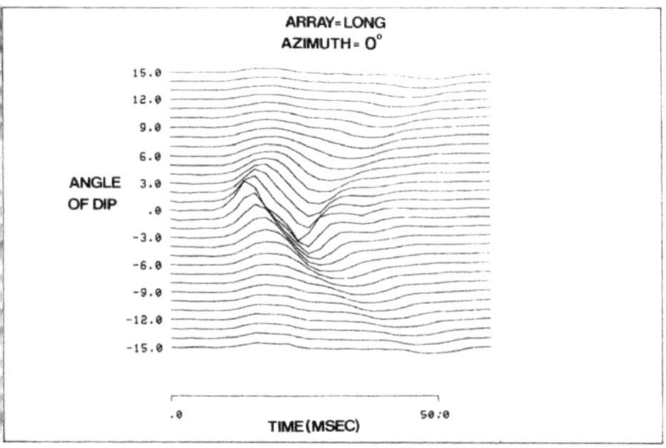

Figure 2.9 :
Emitted signatures for the long array configuration of Figure 2.8. The signatures are plotted at 1° intervals for angles of dip close to the vertical, in the in-line plane.

2.7. DIRECTIVITY OF WIDE ARRAYS

From an operational point of view long arrays are relatively easy to achieve in the field, in that a towed string of sources will naturally align along the direction of motion. For wide arrays this becomes a problem and special equipment must be used. The usual technique is to attach the sources to some sort of rigid float. Using two towing points the float is towed at an angle. This generates lateral lift which moves the float out, wide of the towing line. The lateral position depends on the towing angle and boat speed. Using such techniques, stable wide configurations can be achieved. However, for practical reasons, these wide arrays generally have length as well as breadth, and an overall width of greater than about 100m is difficult to attain.

A typical wide array configuration is shown in Figure 2.10. This array is 70m wide by 70m long and has a tapered shape. The corresponding directivity functions for the in-line and cross-line planes are shown underneath. In contrast to the long configuration of the previous section, this array has geometrical dispersion at all azimuths. The in-line and cross-line directivity functions are therefore quite similar,and less extreme than those of the long array. In particular, the wide array emits a great deal less energy out to the side at high angles of dip. The exact configuration can be adjusted to give an acceptable energy balance. Some of the applications of directivity will be discussed in the next section.

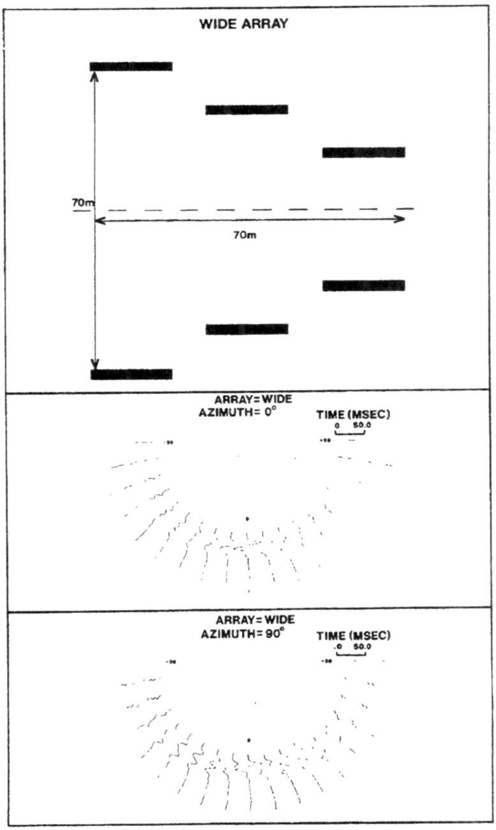

Figure 2.10 :
Directivity functions for a typical wide array. Azimuth = 0° corresponds to the in-line plane, azimuth = 90° to the cross-line plane.

2.8. THE USE OF DIRECTIVITY FOR NOISE SUPPRESSION

The rather different directivity characteristics of long and wide arrays have already been illustrated. In this section the noise problems that can be tackled by careful array design will be considered. It is worth reiterating here that directivity is frequency dependent. Figure 2.5 (c) clearly shows how the beam of emission becomes narrower at higher frequencies. Consequently the effectiveness of beam forming to attenuate noise will also be frequency dependent.

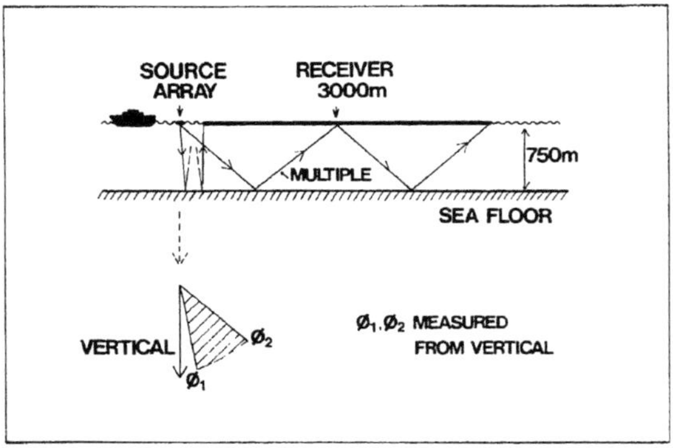

Figure 2.11 :
For the field geometry illustrated, the wedge of angles ϕ_1 to ϕ_2 can contribute to the first order sea floor multiple at the receiver.

As already mentioned the sea surface is a near perfect reflector of seismic waves. In areas where the seabed is also a strong reflector, energy can effectively become trapped, rebounding between sea surface and floor. As a result useful primary seismic data becomes contaminated by multiple images of the sea floor. This 'multiple' energy can be partially suppressed in the data processing or by directivity of the source array. Consider the scenario of Figure 2.11, in which the receiving cable length is 3000m and the water depth is 750m. Energy emitted between the angles ϕ_1 and ϕ_2 can contribute to the first order multiple as shown. ϕ_1 and ϕ_2 are shown plotted against water depth in Figure 2.12. It is clear that if the beam can be restricted to 10°-20°, then a large proportion of the multiple energy will be attenuated for most water depths shown. The source/receiver geometry dictates that received multiple energy will have been emitted close to the in-line plane, at azimuths around 0° . So multiple suppression can be achieved with in-line directivity - in other words by using long arrays. The example above considered the first order multiple. For higher order multiples, the angles ϕ_1 and ϕ_2 become progressively smaller, so progressively tighter in-line directivity is required.

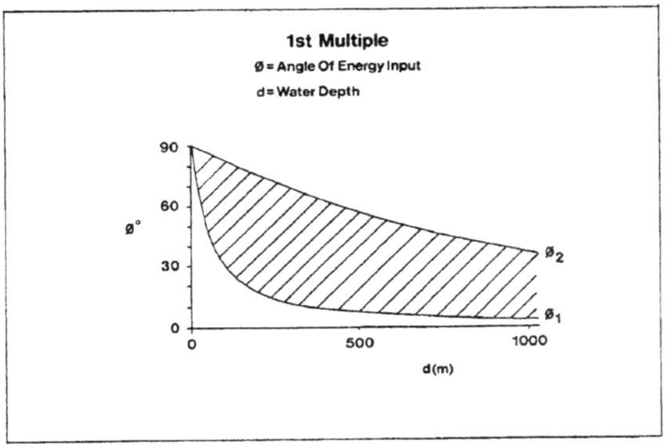

Figure 2.12 :
The range of source angles (ϕ_1 to ϕ_2) which can give rise to the first order multiple, as a function of water depth. This figure assumes a receiver cable length of 3000m.

There are a number of disadvantages to having a very restricted in-line beam of emission :

(1) Reflections from in-line dipping reflectors require energy input at an angle from the vertical, so such reflections will be attenuated.

(2) Reflections from shallow horizontal interfaces will also be attenuated, since these too require energy input at an angle.

(3) The reduction of in-line emission produces enhanced emission at other azimuths (conservation of energy). This can accentuate other noise problems such as sideswipe.

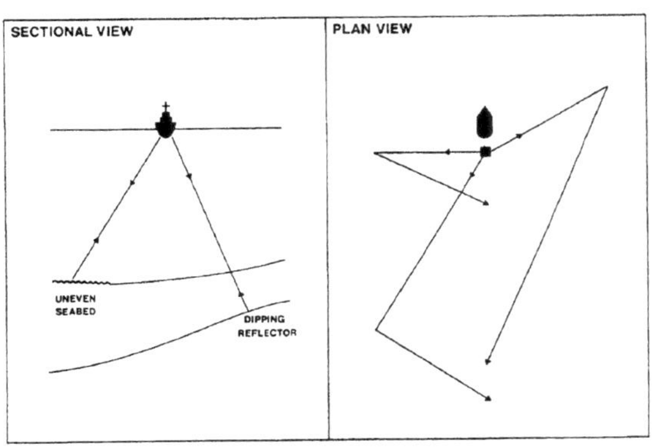

Figure 2.13 :
The phenomenon of sideswipe shown in sectional and plan view. Energy emission at any azimuth can contribute to sideswipe.

The phenomenon of sideswipe is illustrated in Figure 2.13. Energy emitted out to the side can be reflected back from dipping reflectors, or scattered back from an uneven sea floor. Energy emitted at any azimuth can contribute. In fact at small azimuthal angles sideswipe is an important contributor to the primary seismic data set. However as the azimuthal angle increases, the returned data becomes spatially aliased and can be considered as noise. Returns from shallow but distant side scatterers will contaminate even the deeper seismic data. The emission characteristics of wide arrays are suitable for reducing sideswipe, in that the beam of emission is relatively narrow at all azimuths. Figure 2.14 illustrates this property by comparing mainlobe beamwidth versus azimuth for the long and wide arrays of Figures 2.8 and 2.10. Again the principle of conservation of energy is evident in that the reduced wide array emission out to the side is balanced by increased emission around 0° (in-line). In some cicumstances, however excessive reduction of sideswipe can be undesirable. For instance, in 3-D surveys and on strike lines, a degree of sideswipe is required to obtain the necessary information on cross-dipping reflectors.

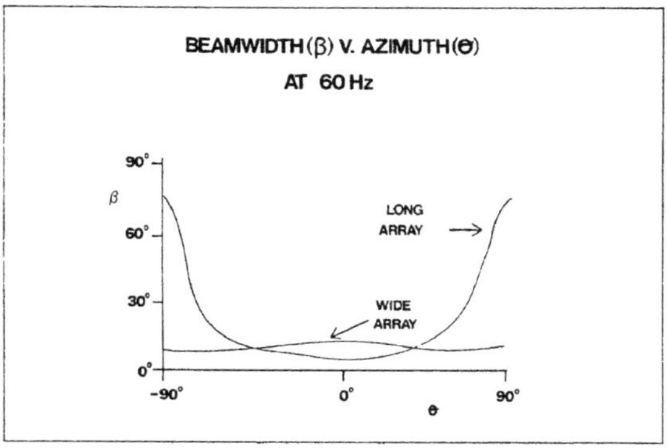

Figure 2.14 :
Mainlobe beamwidth (β) versus azimuth (θ) for typical wide and long array configurations.

Another category of noise that can be attenuated by array directivity falls under the general heading of source generated noise. There is an overlap between this type of noise and sideswipe. Source generated noise embraces direct arrivals from the source, back scattered noise from all azimuths, refracted waves, and other energy resulting from a combination of reflections and refractions in shallow layers. Lynn and Larner (1983) showed that wide arrays were indeed effective at reducing sideswipe and source generated noise. However although the improvement was visible in the raw data from single shots, in stacked data there was no apparent improvement compared to data obtained using compact arrays. They explained this by pointing out that the process of stacking attenuated the same energy as the wide array. Nonetheless in areas in which the side-scattered energy propagates at a velocity similar to the stacking velocity, wide arrays will be effective where stacking will fail.

Considering all the factors discussed above it is clear that there is no universally optimal array design. A thorough understanding of the regional geology is required before specialist long or wide arrays should be used. In large surveys, considerable changes in geology are likely to be encountered, and it is impractical to keep changing array configuration. General purpose designs which have width and breadth, but not too much of each, would appear to be the answer here.

2.9. EFFECTS OF DIRECTIVITY ON PRIMARY DATA

In previous sections it has been shown that directive arrays may be used to attenuate unwanted noise in seismic records. A primary manifestation of this directivity is that the emitted signature varies with direction. If the array is very extended, then the signature can be a sensitive function of direction even within a few degrees of the vertical direction. Now the energy recorded in a seismic shot will have originated from the source at a range of angles, so this signature variation will have a profound effect on the recorded primary seismic data.

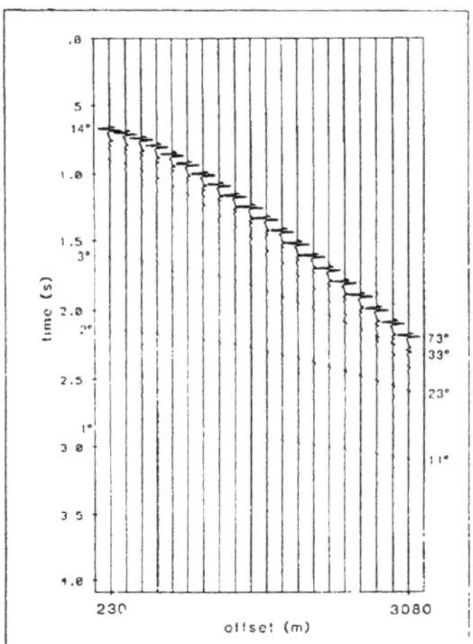

Figure 2.15 :
Synthetic shot file for a point source. The near and far offset source emission angles are marked for the primary events.

Figure 2.15 is a synthetic seismogram for a simple Earth model. A single shot record is shown, the individual traces corresponding to receiver positions 200m apart. In this example a point source was assumed, so the signature shape is constant throughout. The amplitude decay effect is due entirely to spherical spreading. On the left and right hand side of each event the marked angles are the corresponding source emission angles for each position. The angles are furthest from the vertical in the shallow data and for the far offset receiver positions.

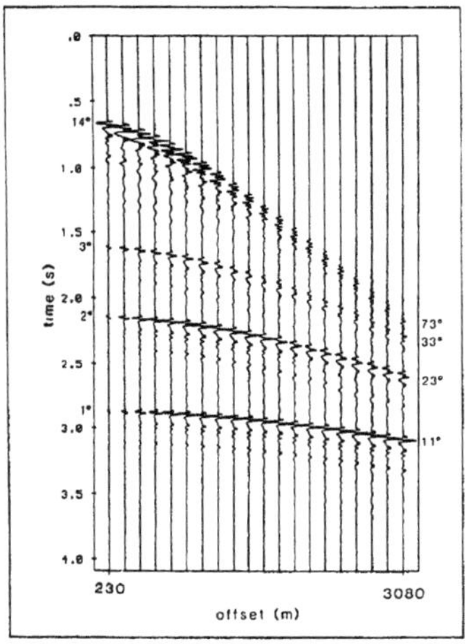

Figure 2.16 :
Synthetic shot file for an array 110m long. The varying signature with angle results from the array directivity.

Figure 2.16 is the same synthetic, this time assuming a directive source array of 110m length. In this and the previous figure the traces have been individually normalised. Nonetheless it is clear that the amplitude behaviour of the events is quite different to the previous example. The signature shape is also quite different particularly in the shallow and at large offsets. Traditional signature deconvolution assumes that there is a single source signature convolved into the data, namely the vertically travelling far field wavelet. This is clearly not the case for extended arrays as demonstrated by Figure 2.16. A deconvolution of this data assuming the single wavelet model is shown in Figure 2.17. As expected this deconvolution breaks down in the shallow and at large offsets. However the severity of this effect is dramatically reduced in fully processed data. For example, the data area where directivity effects are strongest is just the area that is muted out. Similarly, directivity effects are strongest at high frequencies, so the worst data is often filtered out. Loveridge et al. (1984b) showed that for array lengths of up to 100m, 4msec sampling, standard processing and final high cut filters of 60-70 Hz, the effects of directivity in the fully processed data are probably negligible except in the shallowest data. Of course, if a greater bandwidth is required, then the effects will become significant, and directionally dependent deconvolution may have to be considered.

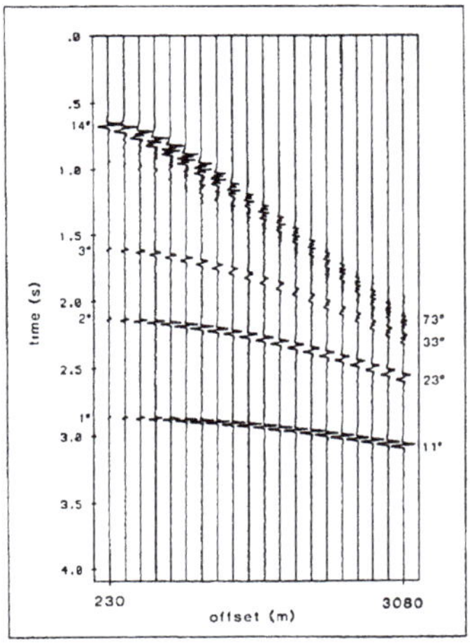

Figure 2.17 : Signature deconvolution of Figure 2.16 using a single filter calculated from the vertically travelling far field wavelet. The limitations of this approximation are clear.

CHAPTER 3

Interaction and Wavefield Determination

The previous chapters have concentrated on more general aspects of seismic sources, such as their physical properties, their variety and the basic principles of directivity. In this chapter, the subtleties of source interaction and some practical aspects of determining signatures in the presence of such interaction will be discussed.

Of course, it should not be forgotten that the real goal of seismic source studies is to improve their deconvolution, thereby increasing the resolution of the seismic method. Historically, deconvolution of such signatures has been the domain of statistical methods as will be discussed in Chapter 5. It is the belief of the authors that such methods have little more to offer. Only by applying deterministic methods can further progress be made. The theme of this chapter is therefore determinism.

3.1. CATEGORIES OF SOURCE INTERACTION

Over the years two distinct categories of interaction have been introduced into descriptions of seismic sources. Regrettably, there has been little consistency of nomenclature with some attendant confusion. The essential distinguishing physical characteristic is whether the 'immediate region of disturbance', for example, the bubble in airgun sources, coalesces with the equivalent for a neighbouring source. This coalescence is highly non-linear and its underlying physics is even today a matter of some debate. In this book, the following terminology will therefore be used:

(a) First order or linear pressure field interaction

This refers to neighbouring sources which do not coalesce. Such sources interact by a coupling of the pressure radiation fields. When arrays of sources started to become commonly used in reflection seismology, it was widely believed that the far field signature of such an array could be synthesized with arbitrary accuracy by the simple act of linear superposition of the signatures of each of the contributing source elements. That this is not the case can be seen clearly in Figure 3.1 which compares the measured far field signature directly beneath a seven-element airgun array with the signature calculated by linearly superposing the seven individual signatures, all measured under controlled conditions. Obvious differences in amplitude can be seen and some indications of phase difference also. The differences become even more pronounced when the same comparison is made after both comparison signatures have been filtered with a 40 Hz high-cut filter as shown in Figure 3.2, a more appropriate bandwidth from the point of view of the processing geophysicist. Again the amplitude differences can be seen but the phase differences have become much more obvious. Note

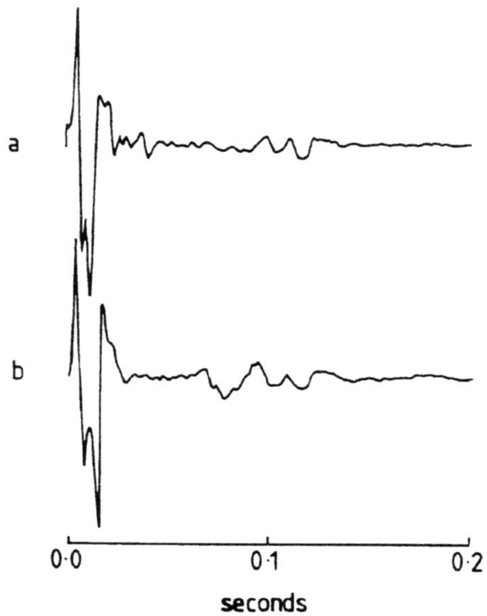

Figure 3.1 :
Comparison between,
(a) signature measured 150m below an array of seven airguns, and
(b) signature computed by superposing the seven signatures obtained by firing each gun in isolation.
The filter setting was out - 360 Hz.

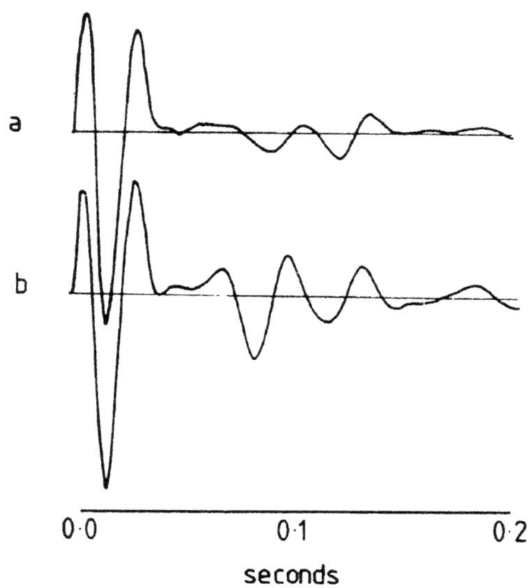

Figure 3.2 :
The same signatures as Figure 3.1, but with a 40 Hz high-cut filter applied. There are obvious differences due to the interaction effects implicit in signature (a).

especially the variation in bubble period. In this case, the physical reason for interaction can be understood in simple terms by considering equation 1.15, which shows that the period of oscillation (amongst other things) of a gas bubble in water depends upon the hydrostatic or water pressure. If the airgun is fired deeper, hydrostatic pressure is greater and hence the period is shorter. Similarly, if the airgun is fired in the proximity of another airgun, then the external pressure it sees is hydrostatic pressure plus a time varying pressure term induced by the radiation of the other source. The period of oscillation and other features of the emitted radiation will therefore change. This effect is here termed linear pressure field interaction.

An important conclusion is that for airgun sources, linear interaction effects are most important at lower frequencies. As has already been discussed however, this frequency dependence is a physical property of the source in use and may be entirely different for another source. To emphasize this point and to understand some of the subtleties of interaction, the following simple model is illuminating.

Consider a system consisting of N springs with weights, wrapped around a circular frictionless cylinder in a damping medium as shown in Figure 3.3. Now consider this system to be set in motion by displacing one or more of the springs from the equilibrium position causing the weights to oscillate. This system is a good model for interaction as the movement of the ith weight is intimately associated with that of the (i-1)th and (i+1)th weights. In other words, they interact.

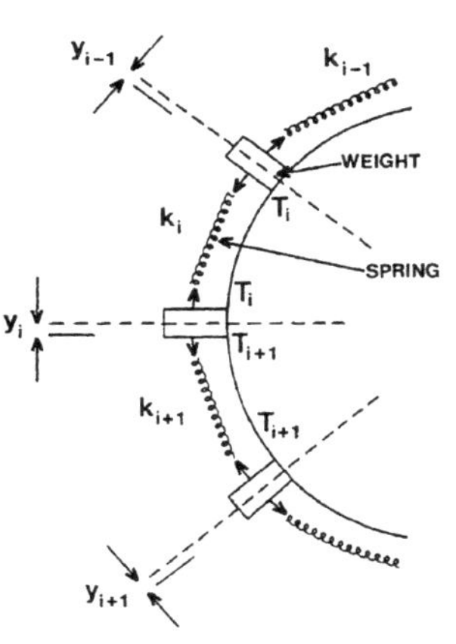

Figure 3.3 :
A plan view of a model for interaction consisting of N springs connected together around a frictionless cylinder.

y_i is the displacement of the ith spring from equilibrium.
T_i is the tension in the ith spring.
k_i is the stiffness of the ith spring.

Strictly speaking, they interact at two levels, one via the springs which does not change with separation and one via the damping medium which obeys the inverse square law. It will be further assumed that the springs are long enough to neglect the latter, although this of course is the normal mode of interaction in the seismic marine environment.

Referring to Figure 3.3, the equation of motion for the ith weight is then

$$\ddot{y}_{\phi_i} = -k_{\phi_i}(y_{\phi_i} - y_{\phi_{i-1}}) + k_{\phi_{i+1}}(y_{\phi_{i+1}} - y_{\phi_i}) - \ell_{\phi_i}\dot{y}_{\phi_i} \tag{3.1}$$

where,

y_{ϕ_i} is the displacement of the ith weight from its equilibrium position.

k_{ϕ_i} is the stiffness of the ith spring.

ℓ_{ϕ_i} is the damping coefficient of the ith weight, dependent on the shape of the weight amongst other things.

ϕ_i = (i modulo N) + 1 because the 1st and Nth springs are connected together.

and $\dot{y}$ denotes dy /dt.

The complete system can be written in matrix form as

$$\begin{bmatrix} \frac{d^2}{dt^2} + (k_1-k_2)+\ell_1\frac{d}{dt} & -k_2 & 0\ \ldots 0 & -k_1 \\ -k_2 & \frac{d^2}{dt^2} + (k_2-k_3)+\ell_2\frac{d}{dt} & -k_3 \quad 0 & 0 \\ 0 & -k_3 & \vdots & \vdots \\ \vdots & 0 & \vdots & \vdots \\ 0 & 0 & \vdots & \vdots \\ -k_1 & 0 & 0 \quad -k_N & \frac{d^2}{dt^2}+(k_N-k_1)+\ell_N\frac{d}{dt} \end{bmatrix} \begin{bmatrix} y_1 \\ \vdots \\ \vdots \\ y_N \end{bmatrix} = \underset{\sim}{0}$$

This system forms a coupled set of ordinary differential equations. In its simplest form, consider N = 2. Then

$$\begin{aligned} \ddot{y}_1 &= -k_1(y_1 - y_2) + k_2(y_2 - y_1) - \ell_1\dot{y}_1 \\ \ddot{y}_2 &= -k_2(y_2 - y_1) + k_1(y_1 - y_2) - \ell_2\dot{y}_2 \end{aligned} \tag{3.2}$$

Combining equations 3.2 yields

$$y_1'''' + (\ell_1 + \ell_2)y_1''' + \ell_1\ell_2 y_1'' + (\ell_2 - \ell_1)(k_1 - k_2)y_1' - \left[(k_1 - k_2) + (k_1 + k_2)^2\right]y_1 = 0 \tag{3.3}$$

Searching for the presence of harmonic solutions suggests substituting

$y_1 = e^{im_1 t}$

for which case, equation 3.3 becomes

$$m_1^4 - i(\ell_1 + \ell_2)m_1^3 - \ell_1\ell_2 m_1^2 + (\ell_2 - \ell_1)(k_1 - k_2)im_1 - \left[(k_1 - k_2) + (k_1 + k_2)^2\right] = 0 \tag{3.4}$$

which gives a general solution of the form

$$e^{i\alpha t}\, e^{\beta t} \tag{3.5}$$

Inspection of these solutions reveals the intricacy of the interacting movement compared with the simplicity of the non-interacting movement of one spring which is simple harmonic motion. Furthermore, the frequency dependence is a function of the physical parameters k_i, l_i.

(b) Second order or non-linear pressure field interaction

Such interaction occurs when the ‘regions of immediate disturbance’ do coalesce. It is fair to say at this stage that the physical understanding of this situation is insufficient to be able to predict exactly what effects this has on the resulting combined signature, and accurate prediction, as described in the remainder of this chapter for linearly interacting sources, is impossible. Be that as it may, non-linear interaction has been deliberately induced by other workers in the past, (c.f. Larner et al. (1982)) as a means of improving primary-to-bubble ratios for airgun source arrays in the absence of a modern understanding of linear interaction. Their rationale depended on the property that two non-linearly interacting airguns appear as a single gun of larger volume and consequently longer period than either of the contributing smaller guns. This allowed them to achieve the broad spread of primary-to-bubble ratios necessary for reasonable statistical deconvolution using Wiener predictive techniques without having to use a similarly broad spread of airgun volumes. In practice, this increases the wear and tear of the non-linearly interacting guns and it is not now widely used. A notable feature of this particular paper is that it studies in some detail the effects of various band-pass filters on the basic source array signature. This is in refreshing contrast to the normal practice of displaying signatures as 'out-out' or without any filtering, under which circumstance the signature is usually nothing like its very band-limited appearance in normal seismic data.

3.2. THE WAVEFIELD OF A LINEARLY INTERACTING ARRAY

The problem of linear interaction between oscillating bubbles of neighbouring airguns has been examined on many occasions with somewhat different conclusions. Ziolkowski (1970) had considered the issue and concluded that it was not a problem provided that at least three bubble diameters separated the interacting bubbles. Giles and Johnston (1973) followed by Nooteboom (1978) continued, with the latter evolving a formula to calculate the minimum non-interacting distance. None of these works however took into account the frequency range of interest.

Further experiments by Lugg (1979) showed that two identical 120 cubic inch (1.97 l) guns just exhibited interaction when separated by a distance of 480 inches (12.2 m). Safar (1976) considered the interaction between identical bubbles using another model based on damped linear oscillators which although not modelling an airgun bubble particularly well did include the important feature that interaction was treated as a modulation of hydrostatic pressure, a fruitful avenue as will be seen. Sinclair and Bhattacharya (1980) although dealing with non-impulsive sources, concluded that interaction is a frequency-dependent phenomenon and that it is a modulation of the hydrostatic pressure field which is responsible.

More recently, Vaage, Ursin and Haugland (1984) reviewed previous attempts to parametrize linear interaction, combining a discussion of the various methods with a careful series of experiments including the effects of a waveshape kit. (A waveshape kit is a baffle introduced into the airgun chamber to reduce the bubble oscillation). Amongst other things they concluded:

(a) Linear interaction effects between airguns should be taken into account in airgun array design.

(b) The primary pulse shape is nearly unaffected by interaction.

(c) Interaction can be used to improve primary to bubble peak-to-peak amplitude ratios although with an accompanying tendency to destabilize the signature under field conditions.

For reference, the minimum distance below which linear interaction takes place is:

Safar (1976)

$$D_s = 6.2\ V_1^{1/3}\ (P_1/P_0)^{1/3.24} \tag{3.6}$$

Nooteboom (1978)

$$D_n = 5.1\ V_1^{1/3}\ (P_1/P_0)^{1/3} \tag{3.7}$$

or assuming isothermal expansion

$$= 8.2\ R_0$$

Johnston (1978)

$$D_j = 2.85\ V_1^{1/3}\ (P_1^{0.341}/P^{0.352}) + d_{gun} \tag{3.8}$$

where,

V_1 = chamber volume of each gun
P_1 = chamber pressure of each gun
P_0 = hydrostatic pressure at airgun depth
d_{gun} = airgun diameter at the exhaust ports
R_0 = equilibrium bubble radius.

(i) Prediction from near field measurements

Building on these results, the problem of linear interaction was first solved by Ziolkowski et al. (1982) with further results presented by the same authors in Parkes et al. (1984b) using airguns as the seismic source, although the method is independent of the physical properties of the source itself. Their method is as follows:

For a single airgun, the oscillations of the bubble are spherically symmetric at seismic frequencies. This is true to at least 500 Hz. The oscillations are driven by the pressure difference between the inside of the bubble and the hydrostatic pressure. The hydrostatic pressure can be taken in the first instance as constant since the rise of the buoyant bubble is comparatively very slow and when stable, will be that of a spherical-cap bubble as can be found in any text-book on fluid dynamics, for example Batchelor (1967). Hence

$$p_d(t) = p(t) - p_H \tag{3.9}$$

where,

$p_d(t)$ is the driving pressure of the bubble,
$p(t)$ is the internal pressure of the bubble and
p_H is the hydrostatic pressure

If $p_d(t) > 0$, it tends to make the bubble expand or slow down collapse. On the other hand, if $p_d(t) < 0$, it tends to make the bubble contract or slow down expansion.

For n guns fired independently, the driving pressure at the ith gun would be

$$p_{d_i}(t) = p_i(t) - p_{H_i} \tag{3.10}$$

Fired simultaneously, this behaviour is modified by interaction. In particular, each gun sees a constant pressure field modulated by a variable pressure field due to the effects of the remaining guns. This may be written as

$$p'_{d_i}(t) = p'_i(t) - p_{H_i}(t) \tag{3.11}$$

where,

$$p_{H_i}(t) = p_{H_i} + m_i(t) \tag{3.12}$$

where $m_i(t)$ is the modulating pressure.

Now equations 3.11 and 3.12 together give

$$p'_{d_i}(t) = \left[p'_i(t) - m_i(t)\right] - p_{H_i} \tag{3.13}$$

Comparing equations 3.13 and 3.10, it may be seen that the modulated case 3.13, can be considered as an unmodulated case 3.10, where the internal pressure is modified to be

$$p'_i(t) - m_i(t) \tag{3.14}$$

Hence, it may be seen that the case of modulating the background pressure field is physically identical to that of modulating the inside pressure of the bubble and keeping the background pressure constant. The important point now is that when the background pressure is constant, linear superposition can be used. Hence, interaction can be modelled by linear superposition of the internally modulated bubbles. Of course, such bubbles do not exist. They are a computational device and will be referred to as 'notional sources'. However, their behaviour can be calculated from a sufficient number of near field measurements.

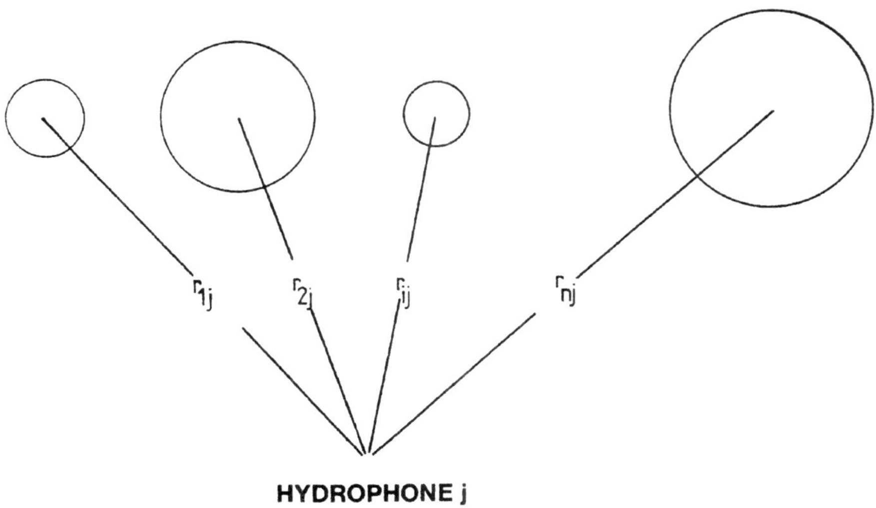

Figure 3.4 : The signature recorded at hydrophone j is the summation of n 'notional' sources, each scaled and delayed in time according to its distance. There will also be a contribution from the n ghost sources.

Consider the geometry of Figure 3.4 and suppose there are n stationary bubbles beneath a free surface with a hydrophone 1m away from each bubble, i.e. there are n hydrophones. Then the voltage output of each hydrophone is given by

$$\frac{h_j(t)}{s_j} = \sum_{i=1}^{n} \frac{1}{r_{ij}} p'_i\left(t - \frac{r_{ij}-1}{c}\right) + \sum_{i=1}^{n} \frac{R}{(r_g)_{ij}} p'_i\left(t - \frac{(r_g)_{ij} - 1}{c}\right) \tag{3.15}$$

where,

$p'_i(t)$ is the ith notional source at 1m.
r_{ij} is the distance from the ith bubble to the jth hydrophone.
$(r_g)_{ij}$ is the distance from the virtual image of the ith bubble to the jth hydrophone.
R is the reflection coefficient at the free surface.
s_j is the sensitivity of the jth hydrophone in volts/bar.

Now if the $h_i(t)$ are measured, the notional sources $p'_j(t)$ can be calculated from equation 3.13 by re-writing it as

$$p'_j(t) = \frac{h_j(t)}{s_j} - \sum_{i=1}^{n(i\neq j)} \frac{1}{r_{ij}} p'_i\left(t - \frac{r_{ij}-1}{c}\right) - \sum_{i=1}^{n} \frac{R}{(r_g)_{ij}} p'_i\left(t - \frac{(r_g)_{ij} - 1}{c}\right) \tag{3.16}$$

and noting that r = 1m for i = j.

Now it can be seen why the hydrophones should be close to each bubble. If this is the case, the first term of the right hand side of equation 3.16, which is the measurement, is an excellent approximation to the notional source allowing a simple iterative solution of 3.16. Note however that it may also be solved directly owing to the nature of the retarded time terms on the right hand side.

In practice, the bubbles are not stationary and Parkes et al. (1984b) modified equations 3.16 by including a constant velocity such that

$$\underset{\sim}{r}_{ij}(t) = \underset{\sim}{r}_{ij}(0) + \underset{\sim}{V}_{ij}t \tag{3.17}$$

where the v and r terms are vectors and v_{ij} is the relative velocity between hydrophones j and bubble i. As they showed, this provides an excellent approximation and acceleration terms, although equally simple to incorporate, are unnecessary. The end product is a predicted far field signature which is arbitrarily close to a far field measurement as is shown by Figure 3.5. It may be concluded that the problem is solved.

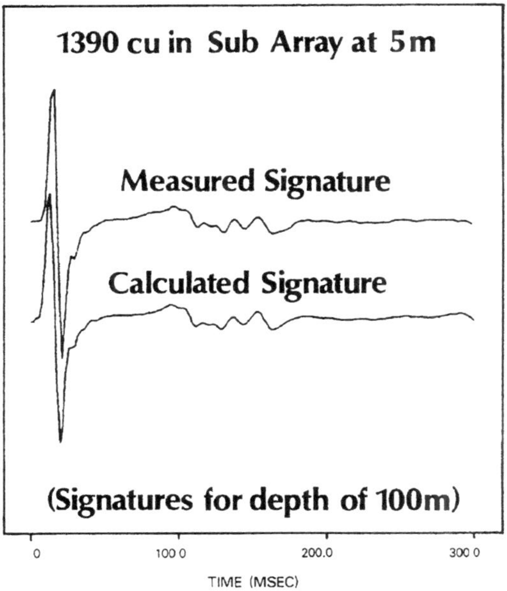

Figure 3.5 :
The top signature is a far field measurement made vertically beneath an array of seven airguns. The bottom signature was calculated from near field measurements using the method described in the text.

It should be noted that if 2n hydrophones are used, the virtual image of the notional sources could also be calculated directly eliminating the need to make any assumptions about the free surface.

In order to use the techniques in practice, it is important to consider operational effects in more detail. First of all, inspection of equation 3.16 shows that the calculation relies on the geometry being precisely maintained and also on a knowledge of the hydrophone sensitivities. Hydrophone sensitivity variation can be accommodated by appropriate calibration. As far as the array geometry is concerned, modern airgun suspension systems based on paravane technology (c.f. Parkes et al. (1984a)) provide excellent stability to the point where the signature changes are negligible over very considerable distance unless pathologic conditions such as the failure of an airgun occur. This is illustrated by Figures 4.2 and 4.3. Figure 4.2 shows a sequence of signatures measured for one sub-array of seven airguns in extreme weather conditions over several km. Figure 4.3 shows a further sequence from the same line and their deconvolution using a single spiking deconvolution filter computed from their average. This stability is in marked contrast to the shot-to-shot variations reported by Hargreaves (1984). This method is considered in more detail next.

(ii) Extrapolation from near field measurements.

An alternative approach to the problem of predicting far field signatures is described by Hargreaves (1984). This method has considerable conceptual elegance and consists of towing a mini-streamer a few metres beneath the source-arrays. A measurement at this mini- or signature streamer can be extrapolated into the far field using wave-field extrapolation techniques based on the Kirchhoff integral solution of the scalar wave equation such that

$$p_{far}(x, y, z, t) = F * \int_{cable} \frac{\cos\theta}{(2\pi r)^{1/2}} p_{cable}\left(x_s, z_0, t - \frac{r}{c}\right) ds \qquad (3.18)$$

where,

F is the inverse transform of $\omega^{1/2} e^{i\pi/4}$,
* denotes convolution.

The author also does a careful analysis of error and concludes that both the random and systematic error of the extrapolation technique are less than the experimental error that arises out of a direct measurement of the far field signature by deep-towed hydrophone.

One restriction of this extrapolation method is that the signature streamer must be in the far field with respect to the cross-line dimension of the array. In practice, this limits it to arrays with such dimensions up to about 20m, which is somewhat less than many operational arrays. As a result of the error analysis, the author concludes that the shot-to-shot variations predicted result primarily from real shot-to-shot variations which, as mentioned earlier, are in marked contrast to the stability reported by Parkes et al. (1984a). It would seem that this highlights deficiencies in the airgun suspension system in use and it would be of interest to see how well this method would fare with a more stable suspension system.

(iii) Scaling Law methods

Another method which may afford a direct treatment of the linear interaction problem is that described in Ziolkowski (1980) and Ziolkowski et al. (1980) whereby scaling laws are used to design arrays of sources in such a way that a deconvolved seismogram may be obtained directly.

In essence, this scaling law states that for two different explosive masses with two different equivalent cavities, the far field wavelets, s'(t) and s(t), associated with the two sources are such that

$$s'(t) = as(t/a) \tag{3.19}$$

where a is the scale factor equal to the cube root of the corresponding ratio of the source masses or source energies. The two seismograms recorded by firing the two sources separately, together with the above scaling law give a set of three equations in three unknowns, enabling a direct solution. Unfortunately, there seem to be fundamental problems with this method as reported for example in the excellent paper by Davies et al. (1984) on the results of the Delft airgun experiment (Ziolkowski (1984a)) which may be due to the non-ideal nature of the source (Vaage, personal communication), and there seems little further to add at this stage until further research clarifies the issues.

(iv) Calibration techniques

This method is due to Newman (1985). In this method, a single secondary source, in this case a watergun, is used to calibrate the array by firing it a short time (about 2 seconds) before the main array. This results in the normal seismogram being the concatenation of two seismograms as shown schematically in Figure 3.6.

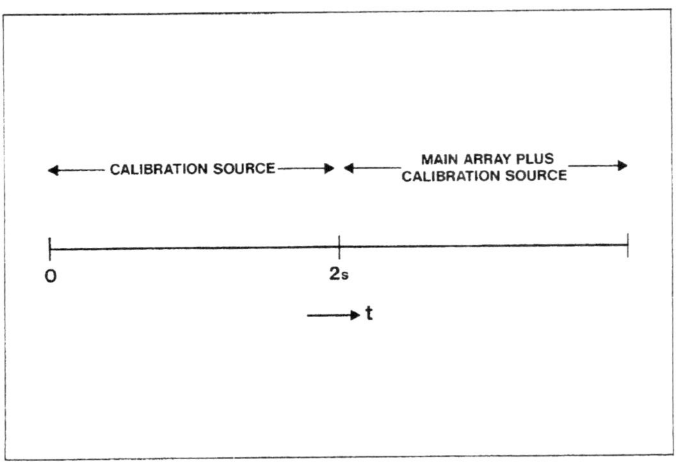

Figure 3.6 : The concatenated seismogram resulting from the firing of a reference source 2 seconds before the main array.

The convolutional model in the first two seconds yields

$$s_r(t) = r(t) * f_r(t) + n_r(t) \tag{3.20}$$

where,

$s_r(t)$ denotes the recorded seismogram
$r(t)$ is the reflection series,
$f_r(t)$ is the far field of the reference source,
$n_r(t)$ is the additive noise appropriate to this seismogram.

After the primary array fires τ seconds later, say, the seismogram yielded is

$$s_p(t) = r(t) + f_p(t) + n_p(t) + s_r(t + \tau) \tag{3.21}$$

The author then assumes that $s_r(t+\tau)$ in equation 3.21 is negligible and indicates various standard techniques for solving equations 3.20 and 3.21 simultaneously to yield the desired result of $f_r(t)$. Unfortunately, these techniques require the standard assumptions about $n_r(t)$ and $n_p(t)$ and their relationship and as the author shows no examples it is difficult to comment further, although it seems to share at least some of the problems of conventional statistical estimation techniques as discussed in Chapter 5. One point made in this work on which the present authors certainly agree is that the trend to source arrays of greater and greater power is highly questionable to say the least.

(v) Numerical modelling.

So far, the theme of the above work is that the far field radiation can be determined in some way from near field measurements. What, however is to be done about the vast amount of data for which no measurements are or were made ? Towards this end, the work of Dragoset (1984) is particularly noteworthy. Building on the original work of Ziolkowski (1970, 1982), the author attempts to predict the signature of an interacting array of airguns based on the physical parameters of each airgun, including the effects of interaction along the lines of Ziolkowski et al. (1982).

The final model for an individual airgun has four important adjustable parameters: the two damping coefficients, the port-open time and the specific heat ratio of air.

By appropriate adjustment of these parameters, quite reasonable agreement between predicted and measured signatures is obtainable, which the author theorises could be further improved by a more realistic model of the behaviour of the gun-port as the gun is fired. Whilst the accuracy of the model is currently inadequate for the demands of signature deconvolution, it provides an admirable method for investigating the gross effects of such parameters as depth and geometry on arbitrary arrays of interacting airguns and contributes substantially to an intuitive understanding of the physics of such arrays.

(vi) Marine Vibroseis

A further step up the ladder of determinism is to force the temporal pressure signature to be of a certain behaviour. On land, this method has been in use with outstanding success for many years and is known as the Vibroseis method. In this method, a plate applied to the medium is moved with a specified temporally varying force known as a chirp in radar technology with a functional form:

$$\begin{aligned} A(t) &= a \sin (bt + ct^2 + d) \\ &= a \sin ([b + ct]\, t + d) \end{aligned} \tag{3.22}$$

which corresponds to a time-variant frequency b + ct, a phase d, and an amplitude a.

Hence A(t) is the source wavelet which can be removed by normal deconvolution techniques, (usually based on correlation in this case). Very recently, this technique has been applied with some success in the marine environment. An added complication here is the fact that the source can move several adjacent ‘shot’ points during the course of the chirp or sweep as it is known. The effects of this on the deconvolution and reduction of lateral resolution in the data are yet to be investigated, but the method has a number of environmental and operational advantages which may compensate. The development of this source in the marine environment will be of considerable interest to reflection seismologists over the next few years.

3.3 ENERGY CONSIDERATIONS

It is worthwhile to discuss the radiated energy and its relationship to directivity and interaction. As has been described in the foregoing, directivity is a geometric effect and interaction pertains to the physical nature of the source in use. It is a property of the scalar wave equation 1.9 that the law of linear superposition is obeyed. In the case of directivity, this means that if p(x,y,z,t) and p'(x,y,z,t) are individually solutions of this equation, then any linear combination of the two solutions is itself a solution. When the sources are fired in such a way as to interact, all that happens is that p and p' <u>change</u> by virtue of this interaction to become p'' and p''' , say, in a manner dependent on the physics of the sources. Neither linear superposition nor the conservation of energy is breaking down here. The problem has simply changed. To expand on this, seismic sources can be categorized in two ways:

(1) Fixed energy sources, e.g. airguns. In these cases a fixed amount of potential energy is imparted to the air bubble at the time it is injected into the water. Thereafter this energy is released, primarily as seismic radiation, although some of it may be lost in heating up the water, etc. The key factor is that the potential energy of the system is fixed at the instant of firing the gun.
(2) Fixed amplitude sources. Consider for example a vibrating plate source in which energy is input into the system throughout the period in which the source emits seismic radiation. The energy available in such a system is not so obviously fixed. Suppose that the system is designed such that the vibrating plate follows a fixed amplitude pattern of vibration, whatever the external pressure. Then in such a system more work must be done, i.e. more energy input, when the external pressure is greater. So, for instance, if the depth of the source varies during its sweep, then the energy input to the source will also vary. For such sources therefore the potential energy of the system is not fixed.

Consider now a single seismic source fired in a medium in which the external pressure is constant with time. From equation 1.14, the total wave energy emitted is proportional to the integral over time and space of the wave pressure squared. For two identical sources, each of which has a fixed potential energy, E, the pressure measured at any point in the far field of each source is P, since the radiation emission is spherically symmetric. In this case, E is simply proportional to the integral of P squared over time and space. Now imagine that the two sources have a horizontal separation of D, and that they are fired together. The law of superposition of wavefields suggests that the combined pressure we would measure vertically beneath the two sources would simply be 2P. We immediately hit a fundamental problem. Suppose that the source separation, D, is vanishingly small, so that the radiation from the two sources together is spherically symmetric, then the energy available appears to be proportional to (2P) squared, i.e. 4E, which is twice the energy available from the individual sources. The problem does not arise when D is large because the apparent quadrupling of energy emission in some directions is balanced by a reduction of energy in other directions in which the waves are out of phase and cancel each other out. Therefore the integrated energy over direction will be 2E, not 4E.

How can the energy conservation paradox apparently resulting when the sources are very close together be resolved ? In these circumstances the two sources are clearly not operating in a medium of constant pressure; rather the external pressure is heavily modulated by the radiation emitted by the other source. At any particular time the pressure may be more or less than hydrostatic pressure, but on average it will be greater. So to maintain an excess pressure wavelet which is identical to the wavelet obtained when the source is fired in isolation, additional energy input to the system is required. Because the external pressure is greater, more work has to be done to increase the pressure by some fixed amount. Now in the case of sources which fall into category (1) above, there is no extra energy available, so the pressure of the emitted wave must reduce to compensate. In fact, the average pressure must be 1/(root 2) less than when the source is fired in isolation. This is the interaction effect. The detailed pressure wavelet emitted by the sources fired together will be quite different to the wavelet emitted by the sources fired in isolation. The form these differences will take in terms of frequency, amplitude and phase will depend entirely upon the physics and dynamics of the source in question. Of course, for sources in category (2) above the situation will be quite different. In these sources the wavelet amplitude is forced by the system, so the interaction effects will force the system to input double the energy, so the energy emitted by two sources will appear to be four times that available from individual sources. Of course, our category (2) sources are idealised and such perfection will be unlikely in practice. Nonetheless the argument illustrates further that the manifestation of interaction effects depends entirely upon the physics of the source in question as was intimated by the springs model introduced earlier.

From the above discussion the following may be concluded:

(a) The source separation at which interaction effects become significant will depend entirely upon the physics of the source. In particular, there is no physical reason to believe that this interaction distance will be a function of frequency.
(b) In fixed energy systems, the energy emitted in the vertical direction by an interacting array may be reduced by up to a factor of 2, compared to the summation of the vertical energy emitted by the same sources fired in isolation. However in energy systems which are not fixed this reduction may be much less.
(c) Arrays may be interacting, or directive, or both. Directivity is entirely a wavefield superposition or geometrical effect, whereas interaction is a physical effect. The two effects are therefore quite different and should not be considered as physically coupled.

3.4. THE AVOIDANCE OF INTERACTION EFFECTS

Given the problems associated with interaction, the reader may have wondered whether it is possible to avoid such complications. A first guess might be that the root of the problem lies in the fact that multi-element sources are used and that if single very powerful sources were used instead, the problem would cease to exist. Unfortunately, quite apart from the operational difficulties associated with the development of such a source, the directivity of multi-element source arrays can be advantageous as has already been discussed in the previous chapter. Perhaps as important however, is the fact that there is no such thing as a point source in use in marine seismology today due to the presence of the virtual image or ghost. It may be called a ghost but as far as the physics of the real source are concerned it is every bit as real and interacts with the real source accordingly although this is a secondary effect unless the source is particularly shallow. Hence if multi-element sources must be used, the only two ways of minimizing primary (i.e. not through the ghost) interaction are:

(a) Avoidance by spatial extension

A simple way of avoiding interaction effects is to spread the sources out laterally so that they are too widely separated to interact. The wavefield can then be determined in any direction by linear superposition of the individual signatures provided of course that they are deep enough so as not to interact with their virtual images in the sea-surface. An unfortunate but inevitable concomitant of spatial spreading is an increase in directivity, which being frequency-dependent, can give a very uneven illumination of reflecting horizons with varying dip, although as was discussed in Chapter 2, directivity can have its uses.

In practice, lateral spreading of sources is done to achieve directivity rather than to avoid interaction.

(b) Avoidance by temporal extension

An intriguing possibility of avoiding interaction by extension whilst simultaneously avoiding unwanted directivity is to spread them out <u>temporally</u>, as described by Stoffa and Ziolkowski (1983). In this method, source n will only be fired when source n-1 has ceased to affect the background pressure field significantly. This again allows linear superposition to be used and appears to work well although the resulting long duration temporal signature, (perhaps several seconds), will lead to some reduction of lateral resolution just as described in Section 3.2. (vi). If the mechanism of interaction was not understood as well as it now is, this would provide an attractive avenue. It is still a reasonable solution in the sense that it requires no additional acquisition hardware and is therefore as reliable operationally as conventional techniques where the signature is parametrized in just two ways; primary-to-bubble ratio and peak-to-peak amplitude.

3.5 SUMMARY

Determinism has a bright future. Currently the most tested method is that descibed in 3.2. (a). In a number of experiments (c.f. Ziolkowski (1984a), Davies et al. (1984)), it appears to be much superior to any known statistical technique and seems the best of the deterministic methods known to date in that the signature is predicted arbitrarily accurately. Competing techniques may eventually achieve the same goal without recourse to near field monitoring and its operational implications.

CHAPTER 4

Practical Aspects of Wavefield Stability

The previous chapters have described how the wavefield of a marine seismic source depends not only on the characteristics of the individual elements of the source but also upon the details of the geometry of the whole source system. Gross effects such as ghosting depend upon depth of deployment, interaction effects depend sensitively upon the relative positions of the elements, and directivity effects are a direct function of geometry. It follows that if the 3-dimensional wavefield of the array is to be stable, then the 3-dimensional geometry of the array must be stable. So geometry plays a major role in wavefield stability, however there are other factors which we can broadly group under the following headings :

(1) The timing synchronisation accuracy of individual elements in an array.
(2) The geometrical stability of the source deployment method.
(3) The intrinsic stability of the acoustic energy generation technique.
(4) The weather.

The above headings are useful for the sake of discussion, however, they are not independent. For instance, geometrical stability will obviously deteriorate with the weather. These topics will be discussed in the remainder of this chapter. Factor (1) is not a problem in practice, factors (2) and (3) vary dramatically from source to source, and factor (4) lies in the laps of the gods!

4.1. TIME SYNCHRONISATION

Correct synchronisation in time is an essential part of a marine source array. The equipment for carrying out this task can be obtained 'off the rack', nonetheless the necessity of synchronisation is so fundamental that it deserves a brief discussion.

The required accuracy of the timing depends on the source signature. As a general rule, the timing errors must be small compared to the duration of the primary radiation pulses of the individual sources. For most sources, the error tolerance is the order of 1 millisecond or less. In most cases, the aim of accurate timing is to align the primary peaks in the emitted wavelet from each array element. Some of the practical issues that the timing equipment must resolve to achieve this are listed below:

(1) The primary radiation peak is not always at the start of the signature (e.g. waterguns).
(2) The time between the start of radiation emission and the onset of the primary peak may not be the same for each array element, even for sources of the same type but different size (e.g. waterguns).

(3) There is usually a time delay between triggering a source, and that source firing. This systematic error may well be different for sources of identical type and size.
(4) The time delay between triggering and firing may drift with time.

Modern timing systems encompass all the effects above in that they detect the true start times of the primary peaks and synchronise accordingly. Continuous monitoring and adjustment corrects for drift.

4.2. SOURCE GEOMETRY STABILITY

In this section some of the practicalities concerning source geometry will be considered. There are two main problems in setting up geometrically stable source arrays at sea. Firstly, the energy sources must be maintained at known and constant depth as they are towed through the water, especially in rough conditions. Achieving and maintaining stable array configurations poses the second main problem. In particular, it is difficult to accomplish stable lateral deviations from the line of tow.

In the majority of marine acquisition systems the seismic sources are hung from independent flotation systems which are towed behind the boat (although some of the less common, more specialist sources are deployed directly from the boat using cranes which overhang the sides). The traditional flotation system consists of Norwegian buoys, one per source. In sub-arrays of buoys, the individual elements are coupled by cables beneath the surface, and the towing points themselves are commonly located on the superstructure near the sources. The main disadvantage of such a system is that the buoys can move independently on the surface and are therefore sensitive to surface waves and swell. The uneven drag on the buoys produced by the waves causes the source suspension cables to oscillate about the vertical. Consequently the source depths also oscillate. The amplitude of this oscillation will depend on the source system and the weather conditions, but a 50 percent depth variation is not unknown !

More recently, a variety of larger, more stable flotation systems have emerged. These are usually long (about 20m), narrow profile floats, which may be flexible or rigid. The rigid type, sometimes called paravanes, are the most stable. The narrow profile of these paravanes permit them to be towed through the water with minimal resistance, cutting through surface waves rather than bobbing up and down as Norwegian buoys are prone to. Furthermore the paravane itself can be towed rather than the sources which hang down freely. These characteristics ensure that the depth (and hence hydrostatic pressure) at each seismic source is maintained accurately. Since emitted waveforms are extremely sensitive to depth, array signature stability can only be achieved with precise depth control. In paravane systems it has been found that even in rough conditions, the gun depths do not vary by more than a few tens of centimetres for source depths of 5-10m (see for instance Parkes et al. (1984a)).

Defining the geometry in the other two dimensions can also present a problem. In particular, obtaining stable lateral deviation from the line of tow requires that the source sub-arrays be equipped with some sort of rudder. In the case of paravanes, the body of the float itself may be used. The principle is illustrated in Figure 4.1. By varying the cable lengths A and B, the amount of lateral 'lift' or Magnus force on the paravane may be adjusted to achieve the desired lateral deviation. This 'wide' deployment method is extremely stable, and lateral deviations of about 50m can be obtained. The actual deviation achieved can be predicted from thin aerofoil theory - see for instance Batchelor (1967).

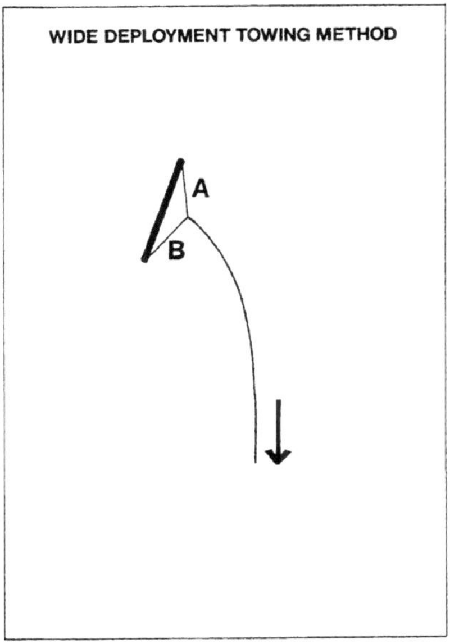

Figure 4.1 :
Source sub-arrays may be deployed with lateral deviations from the line of tow using the method shown. By varying the cable lengths A and B, the amount of lateral lift on the paravane can be adjusted to give the desired lateral deviation.

The radiation emitted by marine seismic sources is in general very sensitive to the external pressure, which in turn is a function of depth. These effects have been studied sytematically and in detail for airgun sources, however there has been very little information published for other source types. From equation 1.15 it is clear that the period of oscillation of an air bubble will get smaller as the depth and hence hydrostatic pressure increases. This effect is non-linear and the exact value depends upon the initial volume and pressure of the gun. As an illustration, for a 100 cubic inch airgun fired at 2000 psi and at a depth of 10m, a 10% variation in depth produces an approximately 3% variation in bubble period. For more details of these effects see for instance Ziolkowski (1971), Ziolkowski (1970), Vaage et al. (1983) and Dragoset (1984).

4.3. SYSTEM STABILITY

Stability considerations must relate to individual systems, so it is difficult to generalise, nonetheless we can illustrate the stability that can be obtained with a system that ranks among the best of those that are available. Figure 4.2 shows signatures of a seven gun sub-array of airguns for a consecutive sequence of about 50 shots. The sub-array float was of the solid, narrow profile type, which appears to be the most stable available. The data were obtained whilst shooting a seismic line in the North Sea, with the airguns deployed at a depth of 5m. At the time the weather conditions were extremely poor (force 8 to 9). The signatures shown in the figure were obtained by directly summing the measurements from seven near field hydrophones which were placed 1m from each airgun. The contribution of the ghost and the second order terms of equations 3.12 and 3.13 have therefore been ignored. Nonetheless the direct summation of the hydrophone measurements is a good measure of signature stability.

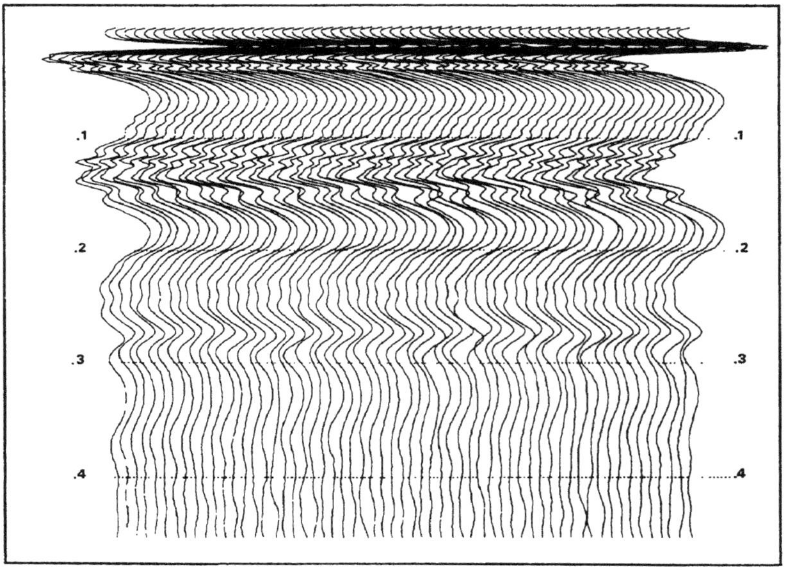

Figure 4.2 : Shot to shot variations of the signature of a 7-gun sub-array of airguns. Even in the bad weather conditions of this test, signature stability is extremely good.

Clearly in this example the signatures are remarkably constant from shot to shot, especially considering the extreme conditions of this test. The most notable shot to shot variations occur at about 1.2 - 1.3 seconds. The similarity of the signatures can be more critically assessed by trying to deconvolve a sequence similar to Figure 4.2 using a single filter. Such an analysis is shown in Figure 4.3. The spiking filter was derived from an average signature calculated from the sequence. The quality of the deconvolution is illustrated on the right. Clearly, this particular source sub-array is extremely stable.

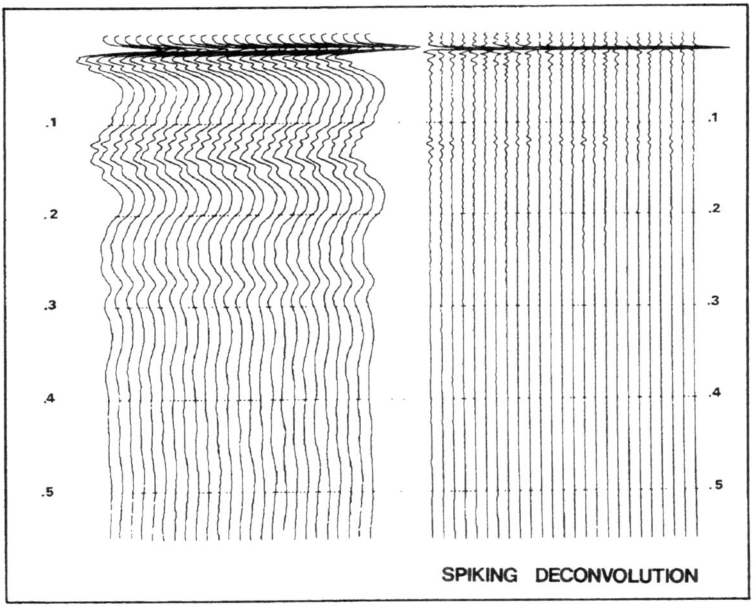

Figure 4.3 : The sequence of signatures on the left have been deconvolved with a single filter to demonstrate their similarity.

The data of the previous two figures were part of a full 10km test line. The correlation of the individual signatures with an average signature over the complete line is illustrated in Figure 4.4. The cross-correlation coefficient is plotted against distance along the line. The remarkable stability of this particular system is once again proven (the perfect case is a constant level of unity). The few mis-fires along the line are only slight, and in fact the largest spike at about 5.5km was caused by a faulty data recording, rather than a source fault.

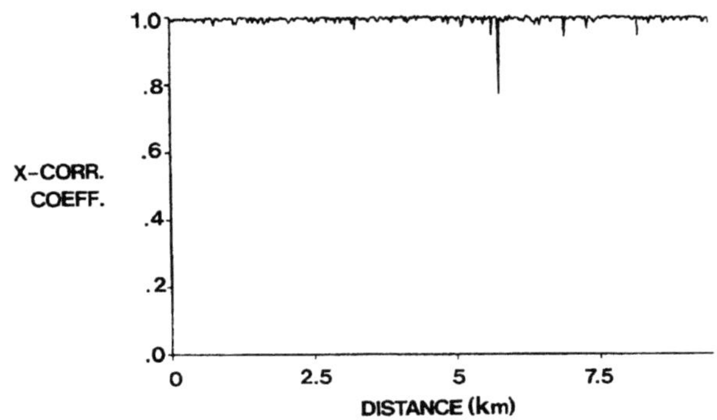

Figure 4.4 : Signature stability in terms of the cross-correlation coefficient between the signature from individual shots along the 10km line and an average for the line.

4.4. THE EFFECTS OF WEATHER

Deterioration in sea state is the main weather effect of concern. Increasing size of sea surface waves and swell has two main effects, (1) source geometry stability deteriorates, and (2) the planar surface approximation that the sea / air interface reflection coefficient = -1 no longer holds. Geometrical stability performance is extremely system dependent. The level of stability that can be achieved in practice has been illustrated in the previous section. Geometry variation will also affect directivity characteristics of arrays. Idealised functions such as those in Chapter 2 will be smoothed out in bad weather. In particular, holes between side-lobes will fill in, and high resolution features will tend to disappear.

In this section the effect of surface waves on the reflection coefficient will be considered. According to Jovanovich et al. (1983), the RMS amplitude of a sea wave distribution is given by equation 4.1 :

$$\sigma_{rms} = \frac{H_{obs}}{2.83} \tag{4.1}$$

where H_{obs} is the observed wave height. This observed height is approximately equal to the average height of the largest one third of the waves (called the significant height).

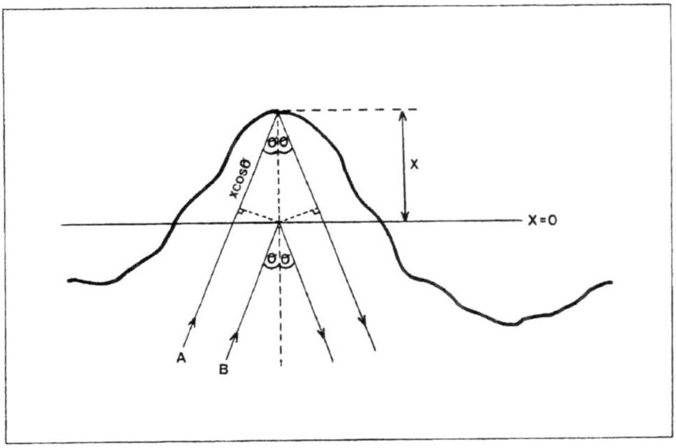

Figure 4.5 : A simple model of a wavy surface in which the path difference between rays A and B equals 2xcos(θ).

The effect a rough surface has on the reflection coefficient itself will be considered by following the formalism of Clay and Medwin (1977). In Figure 4.5 the height of the wave at any point is x, where x=0 is the mean level. Now consider the two rays, A and B, as shown. Ray A reflects from the wavy surface, whereas ray B reflects from a planar surface at the mean level. From the geometry it can be seen that the travel path difference between the two rays is 2xcos(θ). The phase difference, ϕ , is simply the path difference divided by the wavelength and is given by equation 4.2 :

$$\phi = \frac{-2x\nu\cos(\theta)}{c} \tag{4.2}$$

where ν and c are the frequency and velocity respectively. Now the wave elevation, x, will have some distribution, p(x), which will be assumed to be Gaussian and given by equation 4.3 :

$$p(x) = \frac{1}{\sigma(2\pi)^{1/2}} \exp\left(\frac{-x^2}{2\sigma^2}\right) \tag{4.3}$$

where σ is the RMS amplitude of the sea wave distribution given by equation 4.1. The wave reflected from the surface will be the summation over all elevations, x. An average reflection coefficient, R_{av}, can therefore be obtained by integrating the effect of the phase change over all x, as described by equation 4.4 :

$$R_{av} = R_0 \int_{-\infty}^{+\infty} \exp\left(\frac{-2ix\nu\cos(\theta)}{c}\right) \frac{1}{\sigma(2\pi)^{1/2}} \exp\left(\frac{-x^2}{2\sigma^2}\right) dx \tag{4.4}$$

The evaluation of this integral is given in equation 4.5 :

$$R_{av} = R_0 \exp\left(\frac{-2\nu^2\sigma^2\cos(\theta)}{c}\right) \tag{4.5}$$

The average reflection coefficient is therefore a function of angle and frequency, as well as of wave height. Of course, the simplifying assumptions of the above model must be borne in mind. In particular, it assumes that each point on the surface wave acts like a horizontal reflector. The approximation is therefore only valid for small wave-heights and/or long incident wavelengths. For moderate wave heights the relationship should hold for frequencies up to about 250 Hz (see Loveridge (1985)).

4.5. MARINE GHOST VARIABILITY

The variation in the reflection coefficient discussed above will have a profound effect on the wavefields of marine seismic sources. In a vertically travelling far field wavelet, about half the energy results from a coherent sea surface reflection. The effect of a rough surface will be to produce a composite, smoothed out ghost wavelet in the time domain, and in the frequency domain the narrow deep 'ghost notches' will become shallower and broader (see for instance Jovanovich et al. (1983)). An example of this weather effect is shown in Figure 4.6, in which the top signature is a far field measurement of an airgun in good weather conditions, and the bottom signature is a similar measurement in poor weather. The broadening out of the ghost wavelet is quite evident. The high frequencies in the ghost wavelet are removed, so in the frequency domain this 'weather effect' acts like a low pass filter.

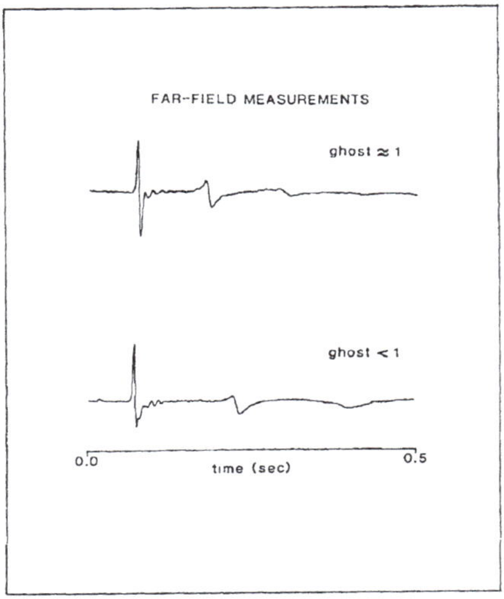

Figure 4.6 :
The top far field measurement was made in good weather and the bottom measurement in poor weather. The reduced ghost size due to surface roughness is evident.

There is another rather different mechanism which may affect the ghost wavelet in certain circumstances (see Loveridge et al. (1984a) and Loveridge (1985)). This mechanism is called the 'shot effect'. Especially with the more powerful sources, it is common to see a surface disturbance immediately above the source at the instant of firing. A fine spray is thrown from the surface which becomes agitated and appears to seeth. For sources such as airguns, this disturbance has nothing to do with the air bubbles which do not reach the surface for several seconds. A particularly demonstrative example of this can be seen in historical pictures of anti-submarine warfare, where the surface is ruptured by a depth-charge almost instantly and several seconds before the expanding gases reach the surface producing a huge fountain of water. The effect is illustrated in Figure 4.7. It can be understood as follows. The atmospheric pressure and the cohesion forces in the water combine to give the water surface a certain tensile strength. Now if the excess pressure in the seismic wave is greater than some critical value, p_c , this breaking strength will be exceeded and the water will be lifted up. A very simple model of this mechanism is illustrated in Figure 4.8. Here it is assumed that all the energy above p_c is removed from the reflected wavelet because it is expended in lifting and vapourising the water and as heat. Therefore any part of the wavelet that exceeds p_c will be clipped at this level.

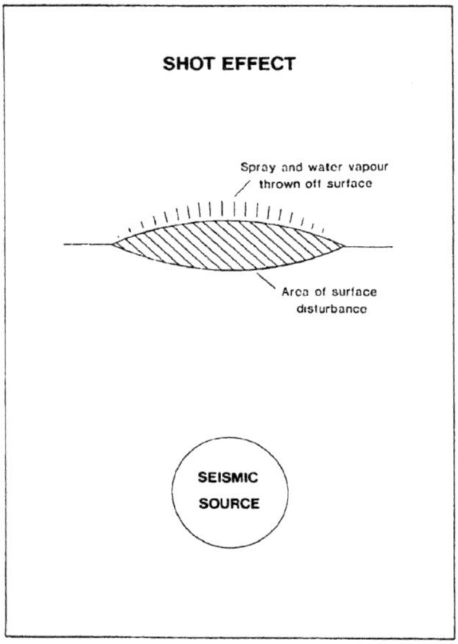

Figure 4.7 :
The 'shot effect'. The radiation from the seismic source produces an area of agitated water and spray at the surface.

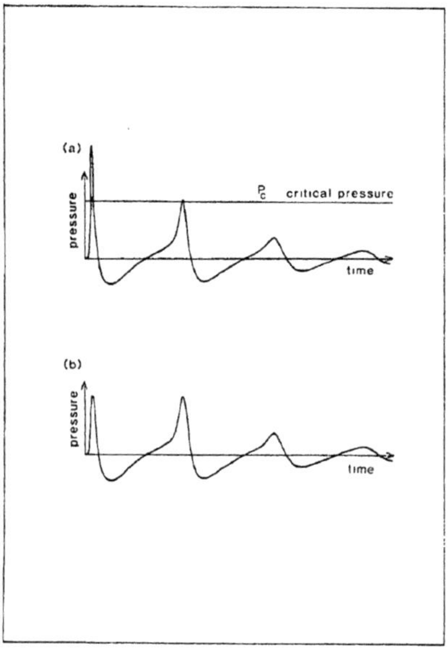

Figure 4.8 :
In a simple model of the shot effect, the peaks of the incident wavelet (a) which exceed some critical pressure, p_c , are removed to give wavelet (b).

The above model is too simplistic in that some of the 'lost' energy will be re-radiated later. There will also be secondary effects in that the roughening of the surface will in itself produce a change in the reflection coefficient in a manner analogous to the weather effect discussed previously. There seems to be very considerable debate as to the exact value of the tensile strength of water, with estimates ranging from less than one bar up to several hundred bars (e.g. Weston (1960), Nyborg et al. (1972)). Nonetheless it is clear that the shot effect does take place. Like the weather effect it will act primarily as a low pass filter and will be most important for very powerful sources and/or for sources which are fired at shallow depths. A full understanding of the mechanism must await detailed experimental and theoretical work.

4.6. SUMMARY

The modern seismic source is extremely robust, predictable and reliable, if a single element is considered under 'laboratory' conditions. However in production acquisition 3-dimensional arrays are towed from boats in the open ocean, and often in far from ideal conditions. In normal operation there are numerous practical problems which must be resolved. The main issues are as follows :

(1) The radiation emission characteristics of the source must be suitable for the job. This is really an issue of source and array design. In general, the emission spectrum must be smooth, without deep notches, and the balance between low and high frequencies must be tuned to the application. As previously mentioned the shape of the time domain wavelet is less important than the spectrum, but must be known with accuracy !

(2) The whole source system must be stable. This includes flotation systems, towing systems and adaptive timing.

(3) The radiation field of the system must be known so that it can be deconvolved from the recorded data.

Point 3 above should cover the possibility that the radiation field will vary from shot to shot. If the system stability is good then shot to shot variations can be minimised. Nonetheless, they will always be present, even in the best of systems. There have been two main approaches to signature deconvolution in the past. One has been to tow a 'far field' hydrophone under the source during acquisition. This goes some way towards including shot to shot variations, but has some major drawbacks, in that the hydrophone is not in the true far field, and it only samples one point in a spatially variant wavefield. The more usual approach has been to measure the vertically travelling far field signature in a well controlled trial, and to use this single signature to deconvolve the wavefield thereafter, assuming that there is no time or space variability of the wavefield. This second technique has become the industry standard, and sometimes works reasonably well, however it has severe and obvious limitations.

Another possibility which has emerged recently has been to use entirely synthetic signatures. Indeed models of some sources are now extremely good and produce signatures that agree very closely with field measurements. Furthermore such synthetics can be used to model the spatial variation of the wavefield. However once again the inability to cope with shot to shot variations is a severe limitation. Nonetheless this method does offer some hope for the reprocessing of old data for which signature measurements were not made.

Perhaps the most promising approach for the future lies with techniques like that described in Chapter 3, in which near field hydrophone measurements are recorded each shot. In theory, such a method allows the full time and space variability of the wavefield to be calculated and later deconvolved. If extra hydrophones are used, then the ghost contribution to the wavefield can also be assessed independently, so that the weather and other ghost effects previously discussed can be included.

A primary aim of the seismic method is to extend the useful bandwidth to higher frequencies and hence improve the spatial resolution of the data. In the final analysis, attaining this goal rests upon achieving increased stability of source systems and/or developing sophisticated techniques for measuring shot to shot wavefield variations and later deconvolving these variations.

CHAPTER 5

Source Signature Deconvolution

5.1. INTRODUCTION

The deconvolution of the seismic source signature has retained the interest of reflection seismologists from the earliest days of the science. Today, with the ever-increasing emphasis on temporal resolution, more is expected of the deconvolution process. The issue of whether such resolution is attainable with the information at hand is unanswered and quite often unaddressed. This is however beyond the scope of this textbook and will not be considered further.

Historically, the merits of the deterministic approach to deconvolution have always been recognized as desirable, but until recently, have not been realized as has already been discussed in Chapter 3. An early noteworthy attempt at signature deconvolution was the Maxipulse* source which achieved considerable success in the early 1970's. This source involved the discharge of a 1/2 lb can of dynamite timed to explode a metre or so away from a signature hydrophone. Variations in the charges themselves necessitated a shot-by-shot deconvolution and similar variation in the shot/hydrophone geometry affected the quality of the result. Nevertheless, the resulting deconvolution was often capable of achieving a signal-to-residual convolution noise ratio of better than 20 db, even though the signature itself was highly oscillatory and in excess of 1 second in duration. The source was eventually displaced by non-explosive sources such as airguns for reasons of economy and ecology as much as quality. However, the deterministic method always places extra demands on an often over-stretched acquisition system and so the growth of statistical methods, whereby 'the' signature was extracted from the recorded data itself, was inevitable.

Before discussing the various merits of the statistical methods, it is worthwhile quickly reviewing just what exactly is the perceived objective of seismic source deconvolution.

Inevitably, the underlying model used is the venerable and highly successful <u>convolutional model</u>, represented by

$$s(t) = \omega(t) * r(t) + n(t) \tag{5.1}$$

where,

$s(t)$ is the recorded seismogram,
$w(t)$ is the wavelet,
$r(t)$ is the earth's reflection series

and

$n(t)$ is additive noise.

* Registered trademark of Western Geophysical

See Ziolkowski (1985b) for a discussion of this model.

In practice, $w(t)$ is itself composed of a number of convolutional responses such as the source and receiver ghosts, the instrument impulse response and so on but these will not be explicitly introduced here.

In an ideal deterministic approach, $w(t)$ is known and a Wiener shaping filter $f(t)$ can be designed such that

$$w(t) * f(t) = d(t) \tag{5.2}$$

where $d(t)$ is the desired output and is usually of short duration and pleasing shape, c.f. Hatton et al. (1986).

Convolving both sides of equation 5.1 with $f(t)$ gives

$$f(t) * s(t) = d(t) * r(t) + f(t) * n(t) \tag{5.3}$$

where the first term on the right hand side corresponds to the desired seismogram and the second term to filtered noise.

Even if $w(t)$ is known, it must be practically invertible, i.e. $N(\nu)/W(\nu)$ is small in some sense, where $W(\nu)$ is the Fourier transform of $w(t)$, $N(\nu)$ is the Fourier transform of the noise and ν is the frequency.

This follows from noting that, in the frequency domain, equation 5.2 can be written

$$F(\nu) = \frac{D(\nu)}{W(\nu)} \tag{5.4}$$

and equation 5.3 can be written

$$F(\nu)S(\nu) = D(\nu)R(\nu) + F(\nu)N(\nu) \tag{5.5}$$

Now, if $F(\nu)\,N(\nu) \gg D(\nu)\,R(\nu)$, noise dominates the inversion. This is equivalent to $N(\nu) \gg W(\nu)\,R(\nu)$. Hence, practical invertibility demands:

(a) A knowledge of $W(\nu)$ for ν in the frequency band of interest.

(b) $N(\nu) \ll W(\nu)\,R(\nu)$, for ν in the frequency band of interest.

In particular, note that notches in the spectrum are only important if the signal-to-noise ratio is poor there, (as it usually is).

5.2 DETERMINISTIC DECONVOLUTION

Having introduced the convolutional model, this section will show some examples of real source signatures and their deconvolution in ideal circumstances. Note that the signatures are shown as the processing geophysicist sees them. This is normally the opposite polarity to that of the acquisition geophysicist.

The first point to make is to re-emphasize the point that when the signature is known by any of the methods described in Chapter 3, for example, the only additional requirement is that it be practically invertible in the sense defined above. The actual temporal shape is irrelevant.

To emphasize this fact, a regal but somewhat extreme source, Windsorseis, is shown in Figure 5.1, with its truly awful amplitude spectrum shown as Figure 5.2. Awful though it may look, this source can still be inverted.

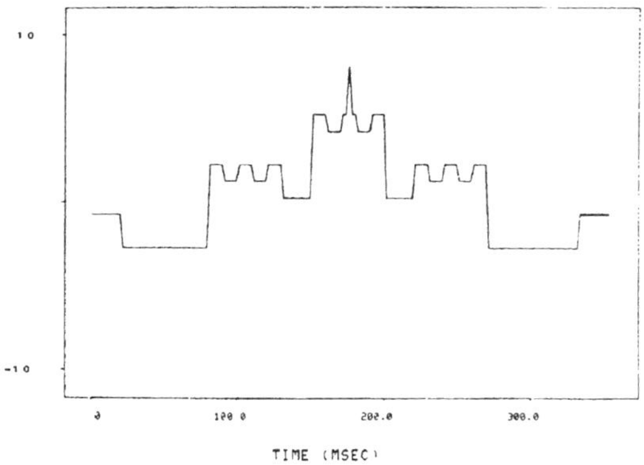

Figure 5.1 :
A hypothetical source, Windsorseis, which has an improbable and extremely undesirable signature (in the classical sense) !

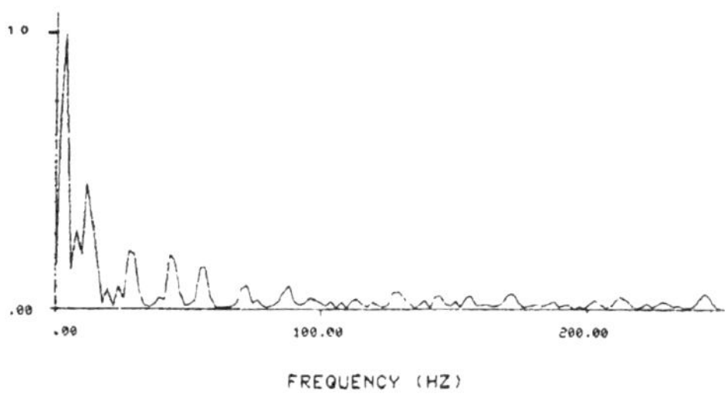

Figure 5.2 :
The amplitude spectrum of the signature of Figure 5.1.

Figures 5.3 and 5.4 show a synthetic reflection series and an actual seismogram computed using this source and the convolutional model, equation 5.1. Using standard Wiener inverse filtering techniques as discussed, for example, by Hatton et al. (1986), the deconvolved seismogram is shown in Figure 5.5. The result is an excellent reconstruction in spite of the pathological nature of the source.

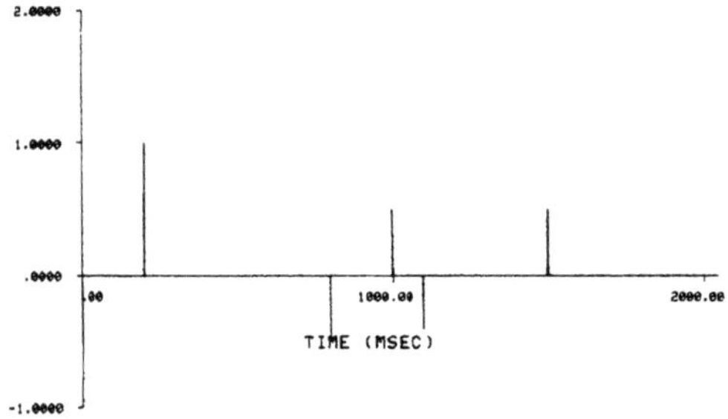

Figure 5.3 :
A synthetic reflection series.

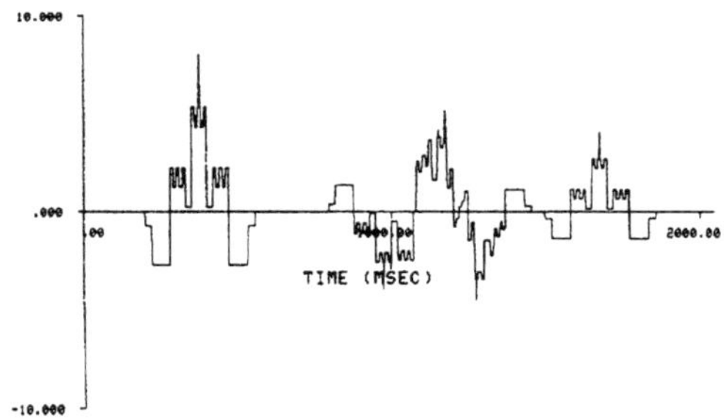

Figure 5.4 :
A seismogram obtained by convolving the reflection series of Figure 5.3 with the source signature of Figure 5.1.

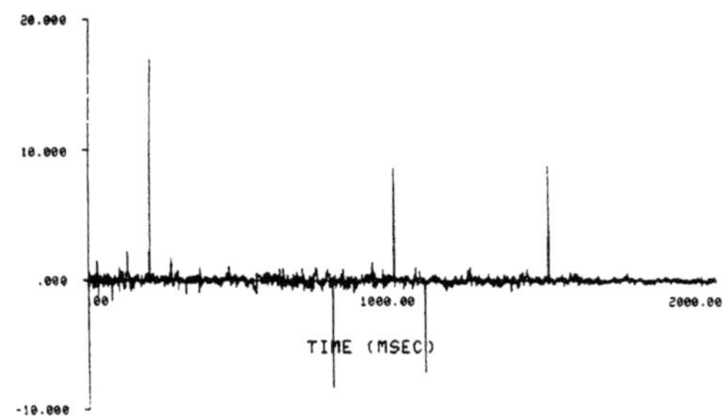

Figure 5.5 :
The seismogram of Figure 5.4 after the source signature has been deconvolved using standard Wiener inverse filtering techniques. Clearly the deconvolved seismogram is an excellent reconstruction of Figure 5.3, despite the pathological nature of the source.

Hence a knowledge of the exact temporal response of the source and a belief in the convolutional model produces a reconstructed reflection series limited only by the effective bandwidth of the source.

Such knowledge is particularly comforting in reality, as the temporal shape of any practical source varies dramatically as a function of bandwidth, amongst other things. Figure 5.6 shows the far field signature of an airgun array of seven airguns with sizes ranging in size from 50 to 305 cu. in., each at a depth of 5m and sampled at 2 msec intervals. Figure 5.7 shows its amplitude spectrum which can be inverted in the approximate bandwidth 4-140 Hz.

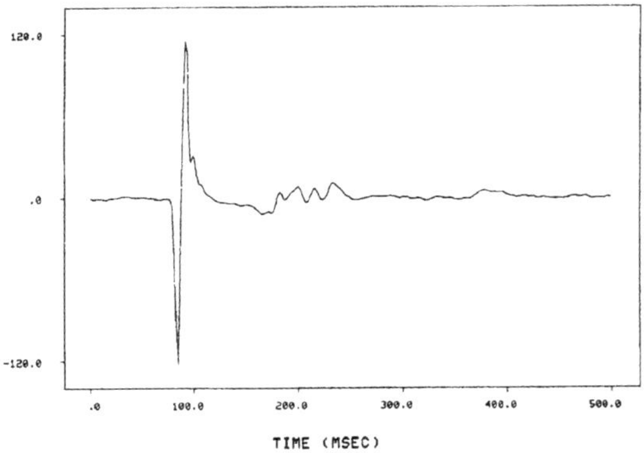

Figure 5.6 :
The vertically travelling far field signature of a typical airgun array with seven guns, deployed at a depth of 5m.

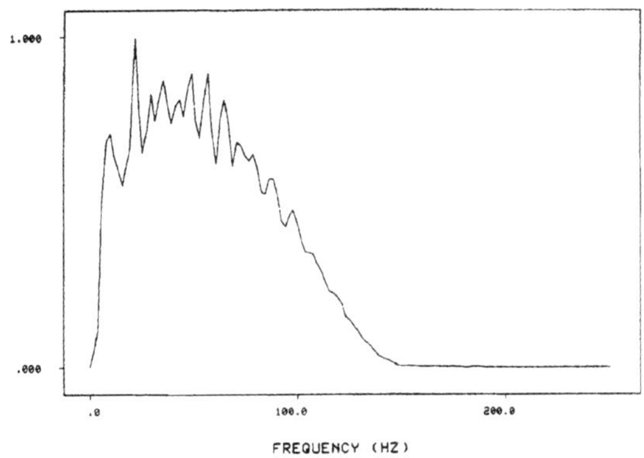

Figure 5.7 :
The amplitude spectrum of the airgun array signature shown in Figure 5.6.

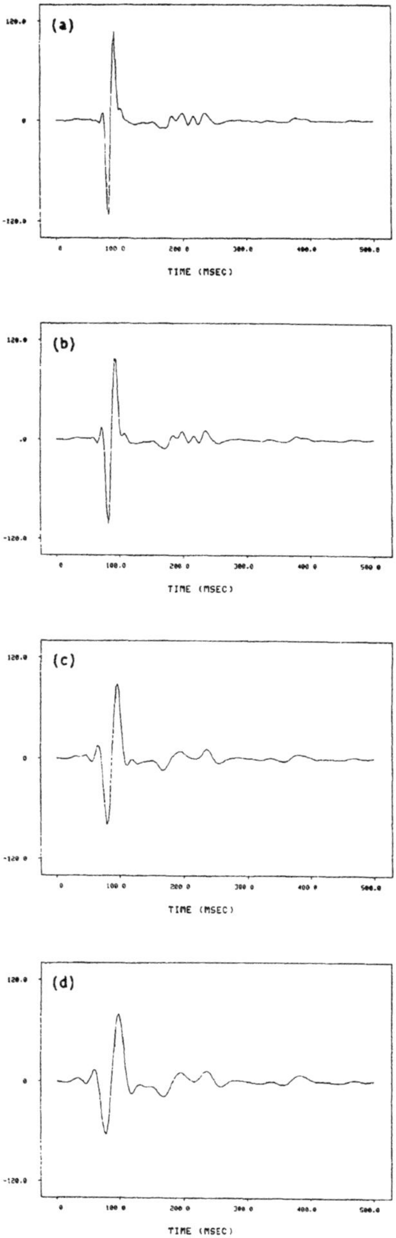

Figure 5.8 :
The signature of Figure 5.6 after filtering with zero-phase bandpass filters as follows,
(a) 10 - 80 Hz
(b) 10 - 60 Hz
(c) 10 - 40 Hz
(d) 8 - 30 Hz.
The filter slopes on the low and high side were respectively 12 and 36 db per octave.

Figures 5.8 a-d show the signature of Figure 5.6 filtered to bandwidths 10-80, 10-60, 10-40 and 8-30 Hz using a zero-phase bandpass filter with slopes of 12 and 36 db per octave on the low and high sides respectively. This models the time-variant filtered appearance such a signature might have on a typical seismic section as two-way travel time increases from 0 to about 6 seconds. The change is considerable and the rapidly decreasing primary-to-bubble ratio shows that the strong low frequency content of this source is associated with the bubble period.

In reality, other effects combine to change the temporal signature even further, particularly absorption. Absorption is often modelled using the assumption that such losses are a constant fraction for each wavelength traversed. This assumption leads to a decay proportional to $\exp(-\pi\nu t/Q)$, where Q depends on the transmitting medium. Figure 5.9 shows the signature of Figure 5.6 after absorption losses corresponding to a Q of 100 at a two-way travel time of 2 seconds, a reasonable value for the North Sea, for example.

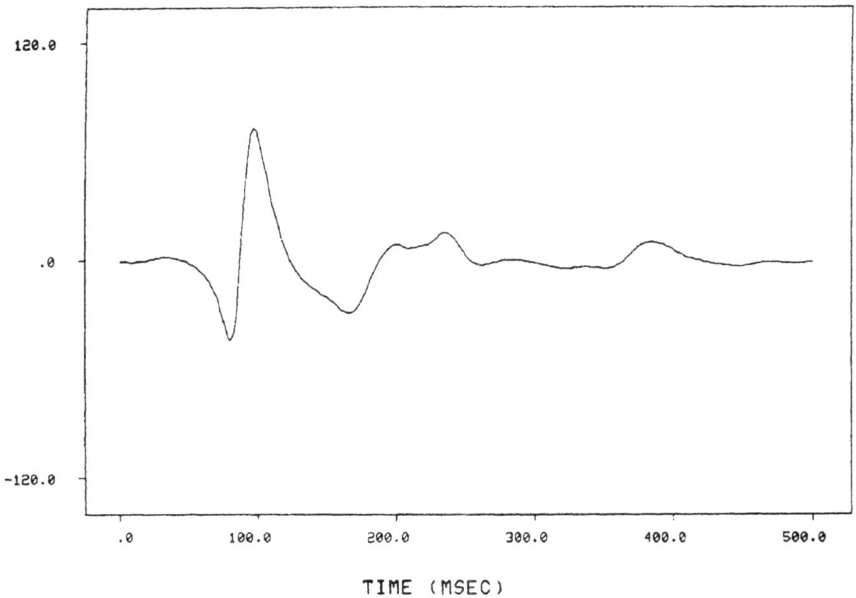

Figure 5.9 : The signature of Figure 5.6 corrected for frequency dependent absorption in the Earth, corresponding to a Q of 100 at a two-way travel time of 2 seconds.

As a comparison with the airgun behaviour described above, Figures 5.10-5.13 show exactly the same suite of examples for a single watergun.

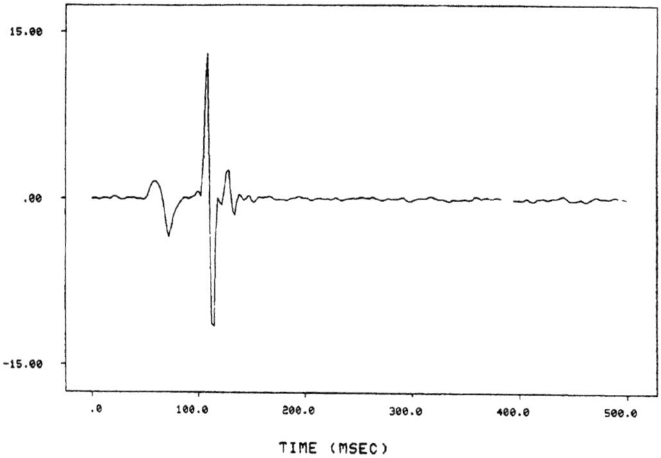

Figure 5.10: The vertically travelling far field signature of a typical watergun array.

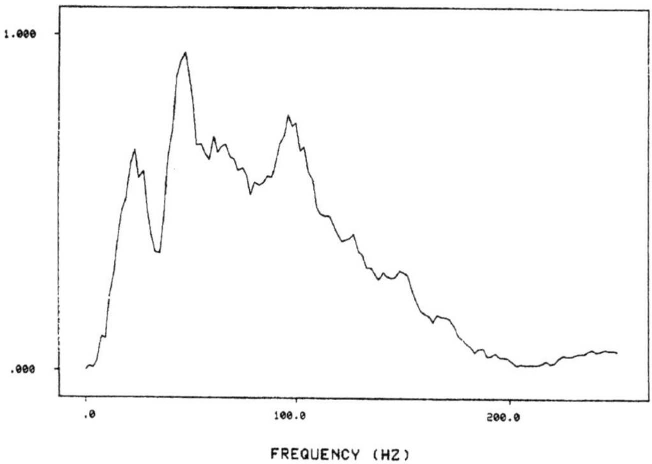

Figure 5.11: The amplitude spectrum of the watergun array signature shown in Figure 5.10.

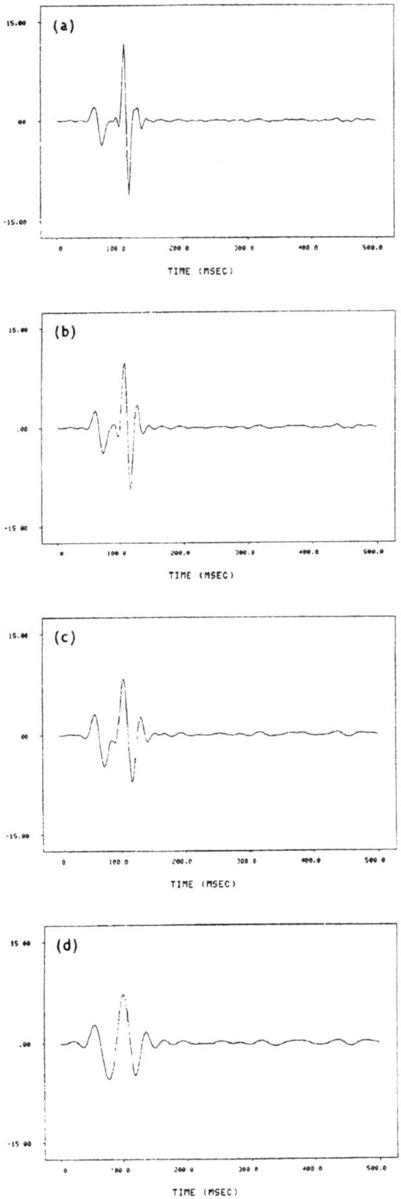

Figure 5.12:
The signature of Figure 5.10 after filtering with zero-phase bandpass filters as follows,
(a) 10 - 80 Hz
(b) 10 - 60 Hz
(c) 10 - 40 Hz
(d) 8 - 30 Hz.
The filter slopes on the low and high side were respectively 12 and 36 db per octave.

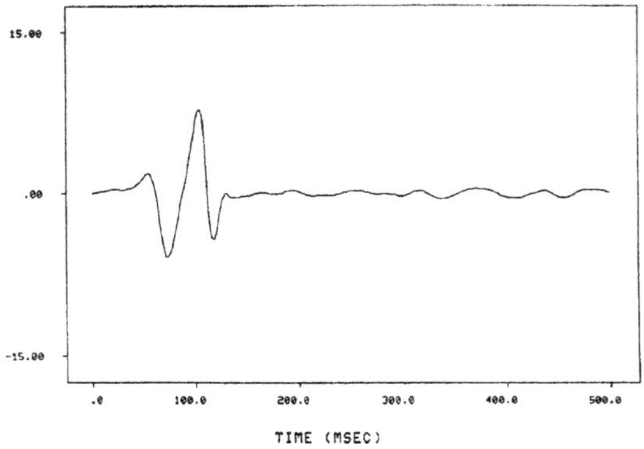

Figure 5.13:
The signature of Figure 5.10 corrected for frequency dependent absorption in the Earth, corresponding to a Q of 100 at a two-way travel time of 2 seconds.

As has been stated previously, both airguns and waterguns are eminently invertible as is shown in Figures 5.14-5.18.

Figure 5.14 shows a typical desired output of a Wiener signature shaping filter.

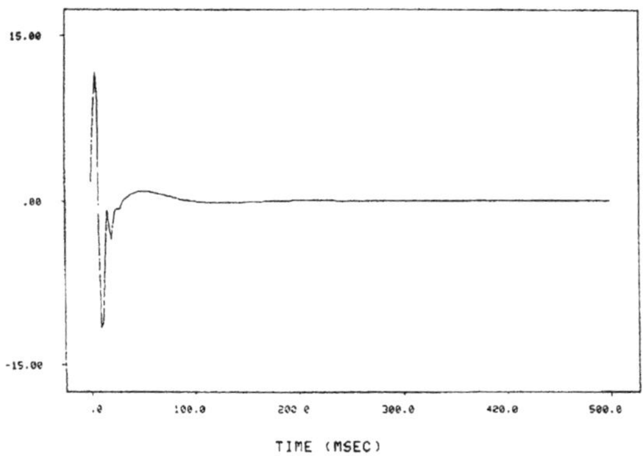

Figure 5.14 :
A Wiener filter is used to transform the source signature into a more 'desirable' wavelet. The wavelet shown is a typical desired output; a minimum phase filter with a passband of 10 - 100 Hz, and low and high slopes of 12 and 36 db per octave respectively.

Figures 5.15 and 5.16 show the derived filters for the airgun array and watergun respectively.

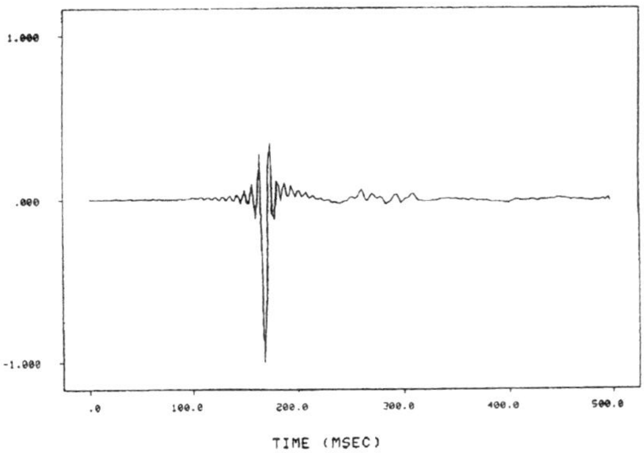

Figure 5.15 :
The Wiener filter that transforms the airgun signature of Figure 5.6 into the desired output of Figure 5.14.

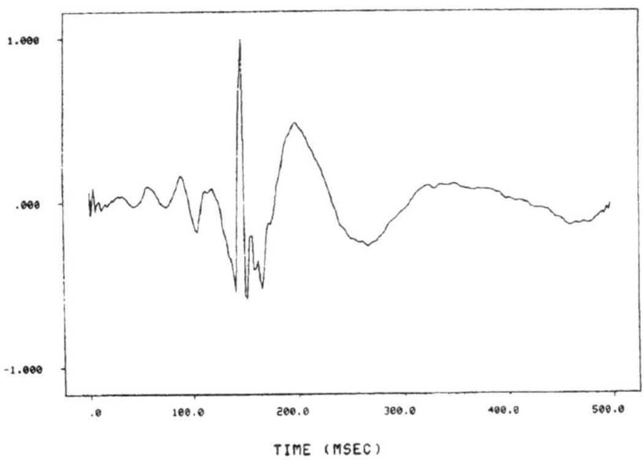

Figure 5.16 :
The Wiener filter that transforms the watergun signature of Figure 5.10 into the desired output of Figure 5.14.

One point worthy of note is that the strong low frequency content of the watergun shaping filter is to compensate for the relative paucity of such frequencies in the watergun source, Figure 5.10, compared with the desired output, Figure 5.14. However, in both cases, an excellent shaping is obtained with a 250 point filter with anticipation component of length 125 points and a white light percentage of 0.2 as can be seen in Figures 5.17 and 5.18, the actual outputs of the Wiener filtering for the airgun array and watergun respectively. As a final comment, note that in this case from Figure 5.15, a considerably shorter filter would have performed just as well for the airgun shaping.

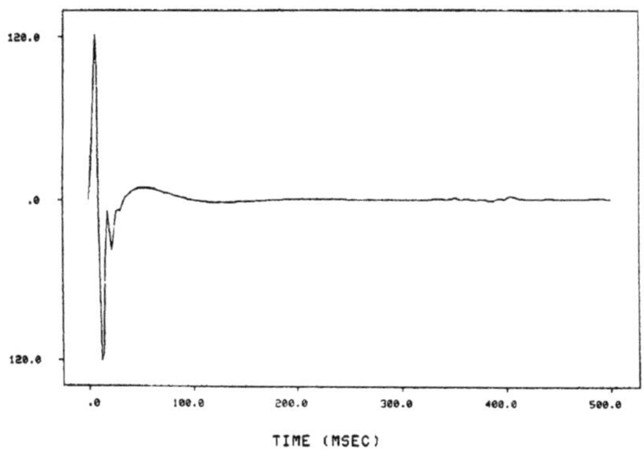

Figure 5.17 :
The resultant shaped airgun array signature (wavelet 5.6 convolved with wavelet 5.15).

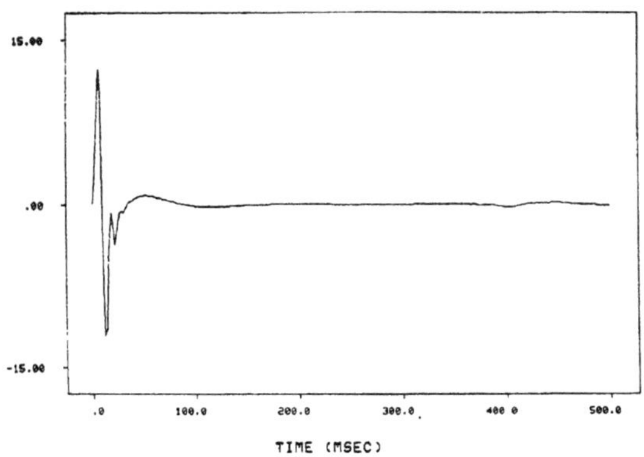

Figure 5.18 :
The resultant shaped watergun signature (wavelet 5.10 convolved with wavelet 5.16).

5.3 STATISTICAL DECONVOLUTION

When looking at the literature, the first thing that impresses is the ingenuity displayed. There are many such methods, each with their own strengths and weaknesses. For more details, the work of Lines and Ulrych (1977) or Jurkevics and Wiggins (1984) should be consulted. Here, only a series of 'thumbnail' sketches will be given, illustrating the most popular statistical methods along with some of their drawbacks.

(i) Predictive deconvolution

This is the standard deconvolution method in seismology, c.f. Peacock and Treitel (1969), and its assumptions have dominated the area of source signature deconvolution for many years. These assumptions are:

(a) Stationarity

(b) Minimum phase condition

(c) Statistically white reflection series.

Under these assumptions, it is possible to reconstruct the wavelet by assuming that the autocorrelation of the recorded seismogram is a good measure of the autocorrelation of the wavelet using (a) and (c) above, and then constructing the unique minimum phase wavelet with this autocorrelation using (b). The classical way of satisfying assumption (b) has been to achieve a primary-to-bubble ratio of some reasonable amount, say 8 to 1 in a broadband recording, and then to state that such a wavelet is minimum phase in the seismic band.

Considering the chain of assumptions, this has served reasonably well. A comparison of this technique against the very accurate notional source deterministic method described in section 3.2 (i) showing its considerable practical limitations can be found in Ziolkowski (1985b). The method can be improved a little by averaging traces from the same shot prior to autocorrelation, but things are unlikely to improve much.

(ii) Forced minimum phase.

This approach has considerable intuitive appeal and is due to Taner (1980). In essence, it is derived from the method described in Section 5.3 (i) with the additional step that the recorded seismogram has an exponential taper applied to it before autocorrelation is applied. After the wavelet is extracted, the inverse of this taper is applied to the resulting wavelet. The rationale is that an exponential taper guards against a mixed phase wavelet in the sense that it forces minimum phase, if a severe enough taper is applied to force all poles of the equivalent z-transform of the wavelet outside the unit circle, c.f.

Robinson and Treitel (1980). The non-minimum phase wavelet is then extracted as in Section 5.3 (i) and the taper is re-applied. The taper is assumed to commute with the process of extraction and therefore the resulting wavelet should be independent of the taper, providing it is severe enough to achieve the desired minimum phase condition. In practice, this is not the case, although it is reasonably satisfactory for wavelets with a small mixed-phase component. See Jurkevics and Wiggins (1984) for a detailed analysis.

(iii) Minimum entropy deconvolution

This method is due to Wiggins (1978). It replaces the assumptions of Section 5.3 (i) with one of sparseness of reflection coefficients in that the recorded seismogram is assumed to be made up of few major changes in acoustic impedance, separated by relatively homogeneous regions. In practice, the method can produce wavelets which are highly variable shot to shot and which are consequently suspect. Again, see Jurkevics and Wiggins (loc.cit.) for more detail.

(iv) Dominant reflection deconvolution

In this approach, a dominant horizon is chosen, perhaps the water-bottom in marine data, and a wavelet extracted directly by gating or windowing the recorded seismogram at the appropriate time. This method can work reasonable in practice, especially if considerable lateral averaging is performed, however, it has the fundamental drawback that the geophysicist is never sure if a dominant horizon is an isolated ‘pure’ reflection or a composite reflection from a number of closely adjacent layers, as is often the case.

(v) Homomorphic methods

These methods derive originally from speech and radar processing. In essence, they work by multiplicative filtering (liftering) of the logarithmic Fourier domain. Although very appealing in theory, they have proved disappointing on real seismic data. Buttkus (1975) and Clayton and Wiggins (1976) show that the real problem is that of unwrapping the phase in excessively noisy data. In very high quality data such as occurs in speech processing, this is achievable. Very high quality seismic data has yet to be acquired.

5.4. PRACTICAL NOTES

Before closing, it is worthwhile discussing another effect which may become important when the major hurdle of determining the temporal response of the source is overcome. As has already been noted, even a single source is directional by virtue of its ghost. Arrays of sources are even more directional. In practical deconvolution, this directionality means that the signature varies according to the direction of emission. Ideally, this geometric effect should be incorporated when deconvolution operators are designed.

As far as statistical methods are concerned, one of the most important contributing factors to their continuing failure is the poor quality of seismic data. Whilst this problem has been diminishing gradually as acquisition methods improve, more is expected of the resulting data by way of compensation. It would seem that until the basic quality improves without correspondingly higher aspirations, the future looks bleak for statistical methods. This is unfortunate as such methods are of course ideal in the sense that no further demands are made on the acquisition system as is not the case for most of the deterministic methods discussed in Chapter 3. However, a lesson learned time and again with seismic data is that progress is made only by considering the acquisition and processing of seismic data as merely different sides of the same coin. With this in mind, deterministic approaches seem to offer the only way forward today.

APPENDIX

Technical Description of Main Sources

There are a large number of different seismic sources presently available and being used. However at the time of going to press 99% of the world production of marine seismic data is using either airguns or waterguns as the seismic source. Technical information on these two sources is therefore given in this appendix.

SOURCE TYPE : AIRGUN

DESCRIPTION : The airgun has become the most widely used of all marine sources because of its simplicity, robustness and reliability. The principle of operation, as the name suggests, is the explosive release of a 'charge' of high pressure air into the water. The traditional design has two air chambers, a control chamber and a firing chamber. The two are connected by a moving shuttle which is kept closed by the pressure of air in the control chamber. At firing time a charge of air is forced under the shuttle pressure plate in the control chamber. The shuttle triggers, opening the firing chamber and venting the air. Some designs have four venting holes, others have a sleeve which allows full 360° venting. On release the bubble of air oscillates producing an essentially periodic decaying signal (see Chapter 1 for more details). The amplitude of the primary pulse and period of oscillation increase with chamber volume. Special modifications to the airguns called waveshape kits are available to attenuate the bubble oscillations if desired.

VARIETIES : Standard pressure, high pressure, sleeve guns, and many variations.

TECHNICAL DATA : Reservoir volume range 0.5 - 10000 cubic inches (more restricted for high pressure and sleeve guns).
Operating pressure:
400 - 5000 psi for standard pressure guns (typically 2000 psi).
4000 - 6000 psi for high pressure guns.
Cycle time typically 3 - 6 seconds.
Energy released per shot 200 - 11000000 Joules.
Weight 18 - 4000 lb.
Primary to bubble ratios of single guns typically 2:1 (0-250 Hz).
Power output examples (peak to zero, 0-250 Hz):
10 cubic inch - 0.5 b-m.
200 cubic inch - 3.0 b-m.

ARRAYS : Airguns are usually fired in arrays in which a mixture of chamber sizes are used to increase the power, increase the primary to bubble ratio, and produce a broad, relatively smooth spectrum. The total airgun volume in typical arrays lies in the range 100 - 8000 cubic inches. The peak to peak output of a 1000 cubic inch array (0-125Hz) would lie in the range 15 - 30 bar-metres, with primary to bubble ratios of about 8:1.

OTHER INFORMATION : Figure A1 shows a representative unfiltered near field airgun signature from a 200 cubic inch gun fired at 2000 psi and 5m depth. Figure A2 shows a representative array signature (vertical far field) from a 910 cubic inch, 7-gun array fired at 2000 psi and 5m depth. Both of the amplitude spectra on the right are normalised and are on a linear amplitude scale.

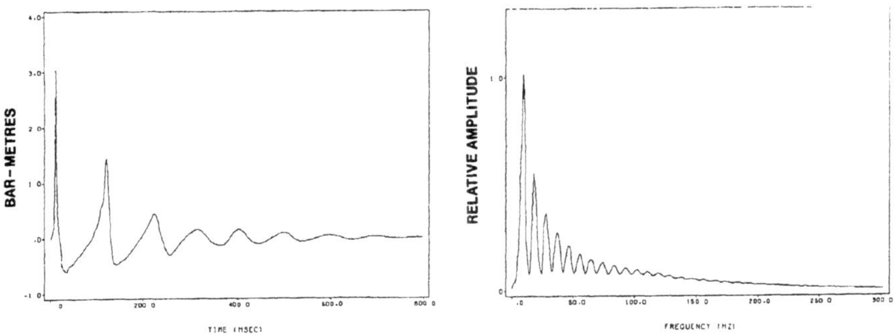

Figure A1 : Near field signature and amplitude spectrum of a 200 cubic inch airgun fired at 5m depth.

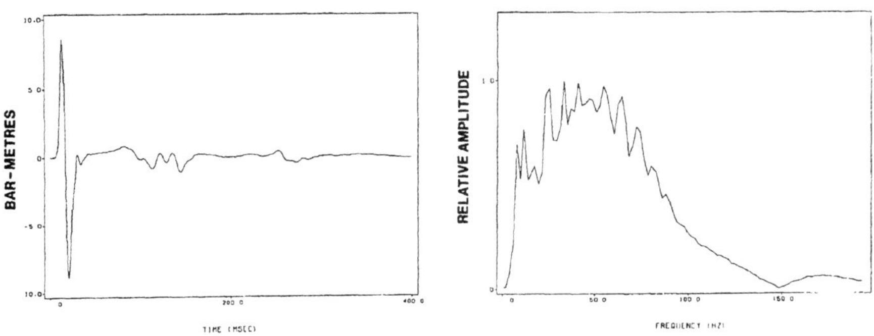

Figure A2 : Vertically travelling far field signature and amplitude spectrum of a 910 cubic inch array of airguns.

REFERENCES : The following list is a selection of papers on airguns. These references and references therein should provide a useful starting point for a more definitive sample of the available literature.

Dragoset (1984)
Giles and Johnson (1973)
Johnson (1978)
Johnson (1980)
Johnson (1982)
Nooteboom (1978)
Parkes et al. (1984a)
Parkes et al. (1984b)
Safar (1980)
Vaage et al. (1983)
Vaage et al. (1984)
Ziolkowski (1970)
Ziolkowski (1971)
Ziolkowski and Metselaar (1984)

SOURCE TYPE : WATERGUN

DESCRIPTION : The watergun is basically an implosive source. In the traditional design there are two chambers, an upper air chamber and a lower water chamber which is open to the sea. A moving shuttle divides the chambers. High pressure air is fed into the top chamber. At firing time the shuttle is released, and the high pressure of the air forces the shuttle downwards. A slug of water is thereby violently ejected from the lower chamber. As the shuttle slows down and the water continues to expand a vacuum cavity is formed between them. The expansion slows, stops, and is followed by an implosion of water into the cavity. The primary peak in the emitted signature is caused by this implosion. This is preceded by a lower frequency pre-cursor produced at the water expansion stage. Since there is no gas bubble, the radiation pulse is short and without oscillations. In general, the emitted radiation is broad band and extends to high frequencies, although some waterguns suffer from deep notches in their spectra at low frequencies.

VARIETIES : Pneumatic, hydraulic and hybrid models available, also combination models which can be operated as airguns or waterguns.

TECHNICAL DATA : Volume range 15 - 500 cubic inches (water emitted).
Water pressure 500 - 3000 psi.
Air pressure 140 - 3000 psi.
Weight 29 - 375 lb.
Some air driven models require air exhaust line to surface.
Firing cycle 0.5 - 8 seconds.
Minimum and maximum firing depths are 1ft to 1800ft.
Power example: a 400 cubic inch gun delivers a peak to zero power of about 3 bar-metres in the 0-125 Hz band, with a primary to secondary ratio of about 4:1.

ARRAYS : Waterguns are frequently deployed in arrays, although unlike airguns it is common for the array elements to be the same size.

OTHER INFORMATION : Figure A3 shows a representative near field signature from a large watergun. Figure A4 shows the vertically travelling far field signature of the same gun. Both of the amplitude spectra on the right are normalised and are on a linear amplitude scale.

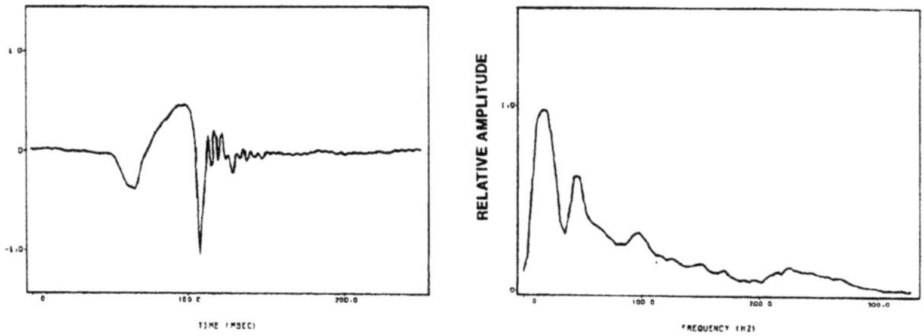

Figure A3 : Near field signature and amplitude spectrum of a large watergun.

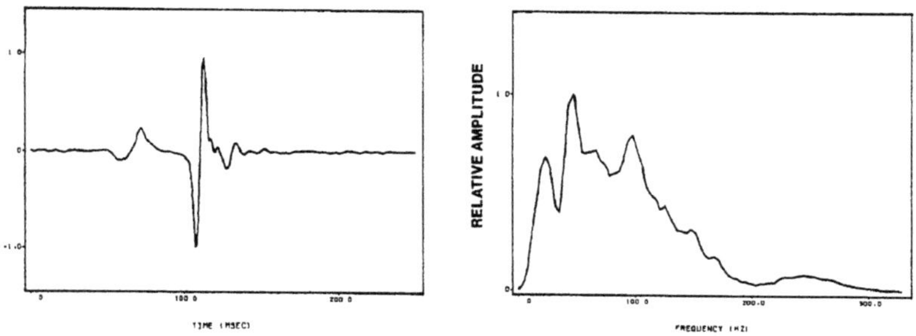

Figure A4 : Vertically travelling far field signature and amplitude spectrum for the watergun of Figure A3.

REFERENCES :

Newman et al. (1977)
Safar (1984)
Safar (1985a)
Safar (1985b)
Tree et al. (1982)

REFERENCES

Aki, K., and Richards, P.G., 1980, Quantitative Seismology - Theory and Methods, Vol. I, W.H. Freeman and Company, San Francisco, U.S.A.

Batchelor, G.K., 1967, An introduction to Fluid Dynamics, Cambridge University Press, Cambridge, England.

Buttkus, B., 1975, Homomorphic filtering - theory and practice, Geophysical Prospecting, v. 23, p. 712-748.

Clay, C.S., and Medwin, H., 1977, Acoustical Oceanography, Wiley Interscience Publications, New York, p. 544.

Clayton, R.W. and Wiggins, R.A., 1976, Source shape estimation and deconvolution of teleseismic bodywaves, Geophys. J. R. astr. Soc. v. 47, p. 151-177.

Coulson, C.A., 1941, Waves, Oliver and Boyd Ltd, Edinburgh, Scotland.

Davies, T.J., Raikes, S.A. and White R.E., 1984, Analysis of some wavelet estimation trials using marine sources, Presented at the 46th annual meeting of the European Association of Exploration Geophysicists, London, England, June 1984.

Dragoset, W.H., 1984, A comprehensive method for evaluating the design of air guns and air gun arrays, Geophysics : The Leading Edge of Exploration, v. 3, No. 10, p. 52-61.

Giles, B.F., and Johnston, R.C., 1973, System approach to air-gun array design, Geophysical Prospecting, v. 21, p. 77-101.

Hargreaves, N.D., 1984, Far-field signatures by wavefield extrapolation, Presented at the 46th annual meeting of the European Association of Exploration Geophysicists, London, England, June 1984.

Hatton, L., Worthington, M.H. and Makin, J., 1986, The theory and practice of seismic data processing, Blackwell's, Oxford, England.

Johnston, R.C., 1978, The performance of marine airgun arrays of various lengths and sizes, 48th S.E.G. meeting, San Francisco, California.

Johnston, R.C., 1980, Performance of 2000 and 6000 psi air guns : theory and experiment, Geophysical Prospecting, v. 28, p. 700-715.

Johnston, R.C., 1982, Development of more efficient airgun arrays : theory and experiment, Geophysical Prospecting, v. 30, p. 752-773.

Jovanovich, D.B., Sumner, D., and Akins-Easterlin, S.H., 1983, Ghosting and marine signature deconvolution : a prerequisite for detailed seismic interpretation, Geophysics, v. 48, p. 1468-1485.

Jurkevics, A., and Wiggins, R.A., 1984, A critique of seismic deconvolution methods, Geophysics, v. 49, p. 2109-2116.

Larner, K.L., Hale, D., Misener Zinkham, S. and Hewlett, C.J.M., 1982, Desired seismic characteristics of an airgun source, Geophysics, v. 47, p. 1273-1284.

Lines, L.R. and Ulrych, T.J., 1977, The old and the new ain seismic deconvolution and wavelet estimation, Geophysical Prospecting, v. 25, p. 512-540.

Loveridge, M.M., 1985, Marine Seismic Source Signatures, Directivity and the Ghost, D. Phil. thesis, University of Oxford.

Loveridge, M.M., Parkes, G.E., and Hatton, L., 1984a, A study of the marine ghost, Presented at the 46th Annual Meeting of the European Association of Exploration Geophysicists, London, England, June 1984.

Loveridge, M.M, Parkes, G.E., Hatton, L., and Worthington, M.H., 1984b, Effects of marine source array directivity on seismic data and source signature deconvolution, First Break, v. 2, No. 7, p. 16-22.

Lugg, R., 1979, Marine seismic sources, in Developments in geophysical exploration methods, Ed. A.A. Fitch, Appl. Sci. Publ., London.

Lynn, W., and Larner, K., 1983, Effectiveness of wide marine seismic source arrays, Presented at the 45th Annual Meeting of the European Association of Exploration Geophysicists, Oslo, Norway, May 1983.

Newman, P., 1985, Continuous calibration of marine seismic sources, Geophysical Prospecting, v. 33, p. 224-232.

Nooteboom, J.J., 1978, Signature and amplitude of linear airgun arrays, Geophysical Prospecting, v. 26, p. 194-201.

Nyborg, W.L., Scott, A.F., and Ayres, F.D., 1972, Tensile strength and surface tension of liquids, in American Institute of Physics Handbook, 3rd. edition, Ed. Gray, D.E., Mcgraw-Hill, New York, p. 2312.

Parkes, G.E., Hatton, L., and Haugland, T., 1984a, Marine source array directivity : a new wide airgun array system, First Break, v. 2, No. 7, p. 9-15.

Parkes, G.E., Ziolkowski, A., Hatton, L., and Haugland, T., 1984b, The signature of an air gun array : computation from near-field measurements including interactions - practical considerations, Geophysics, v. 49, p. 105-111.

Peacock, K.L. and Treitel, S., 1969, Predictive deconvolution: Theory and practice, Geophysics, v. 34, p. 155-.

Newman, P., Haskey, P., Small, J.O., and Waites, J.D., 1977, Theory and application of water gun arrays in marine seismic exploration, Presented at the 47nd Annual International Meeting of the Society of Exploration Geophysicists, Calgary, Canada, October 1977.

Peacock, J.H., Peardon, L.G., Lerwill, W.E., and Wisotsky, S., 1982, An evaluation of a wide-band marine vibratory system, Presented at the 52nd Annual International Meeting of the Society of Exploration Geophysicists, Dallas, USA, October 1982.

Robinson, E. and Treitel, S., 1980, Geophysical Signal Analysis, Prentice-Hall, New Jersey, U.S.A.

Safar, M.H., 1976, The radiation of acoustic waves from an airgun, Geophysical Prospecting, v. 24, p. 756-772.

Safar, M.H., 1980, An efficient method of operating the air-gun, Geophysical Prospecting, v. 28, p. 85-94.

Safar, M.H., 1984, On the S80 and P400 water guns : a performance comparison, First Break, v. 2, No. 2, p. 20-24.

Safar, M.H., 1985a, On the calibration of the water gun pressure signature, Geophysical Prospecting, v. 33, p. 97-109.

Safar, M.H., 1985b, Single water gun far field pressure signatures estimated from near-field measurements, Geophysics, v. 50, No. 2, p. 257-261.

Sinclair, J.E. and Bhattacharya, G., 1980, Interaction effects in marine seismic source arrays, Geophysical Prospecting, v. 28, p. 323-332.

Stoffa, P.L. and Ziolkowski, A.M., 1983, Seismic source decomposition, Geophysics, v. 48, p. 1-11.

Taner, M.T., 1980, Wavelet processing through all stages of seismic data processing, 42nd E.A.E.G. meeting, Istanbul, Turkey.

Tree, E.L., Lugg, R.D., and Brummitt, J.G., 1982, Why water guns ?, Presented at the 52nd Annual International Meeting of the Society of Exploration Geophysicists, Dallas, USA, October 1982.

Vaage, S., Haugland, K., and Utheim, T., 1983, Signatures from single airguns, Geophysical Prospecting, v. 31, p. 87-97.

Vaage, S., Ursin, B., and Haugland K., 1984, Interaction between airguns, Geophysical Prospecting, v. 32, P. 676-689.

Weston, D.E., 1960, Underwater explosions as acoustic sources, Proc. Phys. Soc., v. 76, p. 233-249.

Wiggins, R.A., 1978, Minimum entropy deconvolution, Geoexpl., v. 16, p.21-.

Ziolkowski, A., 1970, A method for calculating the output waveform from an air gun, Geophys. J. R. astr. Soc., v. 21, p. 137-161.

Ziolkowski, A., 1971, Design of a marine seismic reflection profiling system using air guns as a sound source, Geophys. J. R. astr. Soc., v. 23, p. 499-530.

Ziolkowski, A.M., Lerwill, W.E., March, D.W., and Peardon, L.G., 1980, Wavelet deconvolution using a source scaling law, Geophysical Prospecting, v. 28, p. 872-901.

Ziolkowski, A.M., 1980, Source array scaling for wavelet deconvolution, Geophysical Prospecting, v. 28, p. 902-918.

Ziolkowski, A., 1982, An airgun model which includes heat transfer and bubble interactions, 52nd Annual S.E.G. meeting, Dallas, Texas.

Ziolkowski, A., Parkes G.E., Hatton L. and Haugland T., 1982, The signature of an air gun array: Computation from near-field measurements including interactions, Geophysics, v. 47, p. 1413-1421.

Ziolkowski, A., and Metselaar, G., 1984, The pressure wavefield of an airgun array, Presented at the 46th Annual Meeting of the European Association of Exploration Geophysicists, London, England, June 1984.

Ziolkowski, A., 1984a, The Delft airgun experiment, First Break, v. 2, p. 9-18.

Ziolkowski, A., 1984b, Deconvolution, IHRDC, Boston.

INDEX

Absorption 87
Acoustic wave generation 3
Acoustic wave propagation 4
Airgun array 85
Airgun(s) 3, 47, 50, 58, 60, 100
Arrays 25
- long 33
- tuned 25
- wide 34
Average reflection coefficient 74

Bandwidth 45
- effective 85
Bar 19
Bar-metres 20
Beam forming 36
Beam-steering 25, 31
Bubble motion 8
Bubble period 69, 87

Chirp 59
Coalescence 45
Convolutional model 21, 58, 81

Determinism 45, 62
Deterministic deconvolution 83
Deconvolution
- directionally dependent 41
Directivity
- effects of 39
Directivity functions 29
Dominant horizon 94
Dominant reflection deconvolution 94

Effective bandwidth, see Bandwidth
Effects of directivity see Directivity
Elasticity
- bulk modulus of 6
Elastic waves 4

Energy 7, 29
- conservation of 8
- - law of 28
Energy conservation paradox 60
Equation of motion 5
Exponential taper 93

'Far field' 15
'Far field' hydrophone 77
'Far field' measurements 16
'Far field' signature 10, 55
Flotation 68
Forced minimum phase 93
Fringe patterns 27

Ghost 10
Ghost notches 12, 13, 74

Interaction 45, 58, 61
- linear 45, 50
- non-linear 49
Interaction distance 51
Interference 12, 28
Isentropic 5

Law of conservation of energy, see Energy
Law of superposition of wavefields, see Superposition
Linear interaction, see Interaction
Linear relative velocity model 18
Linear superposition, see Superposition
Long arrays, see Arrays

Mainlobe 38
Marine ghost variability 74
Marine vibroseis 59
Mass equation
- conservation of 4

Measurement of signatures 16
Minimum entropy deconvolution 94
Mis-fires 71
Multiple images 36

Navier-Stokes equation of motion 5
Near field 15
- measurements 18, 53
Noise suppression 36
Non-linear interaction, see Interaction
Norwegian buoys 68
Notional source(s) 17, 53, 54
Numerical modelling 58

Paravane 68
Paravane technology 55
Particle velocity 15
'Peak to peak amplitude' 19
Period of oscillation 9
Phase delay 27
Point sources 16
Predictive deconvolution 93
Pressure 7
Primary to bubble ratio(s) 19, 20, 50, 87, 93
Pulse duration 20
Pulse repeatability 20, 21
Pulse shaping 25

Reflection coefficient 73
Relative motion 17
Rough surface 73

Scaling law 56
Sea wave distribution 72
Seismic source deconvolution 81
Shot effect 75
Shot to shot variations 71, 77
Sidelobes 28, 31, 72
Signature deconvolution 81, 93
Significant height 73
Sideswipe 37, 38
Sound velocities 7
Source generated noise 38
Sparker 4
Spatial extension 62
Speed of sound 6
Spherical-cap bubble 51
Spherical spreading 40
Spherical wave equation 6
Stability 68
Statistical deconvolution 93
Superposition
- law of 60
- - of wavefields 17
- linear 53, 62
System stability 70

Temporal extension 62
Time synchronisation 67
Timing errors 67
Total energy emission 8
Tuned arrays, see Arrays

Velocity 7
Virtual image 10

Watergun(s) 3, 57, 67, 88, 103
Wave height 73
Waveshape kit 50
Weather 55, 67, 72, 74
- effect 74
Wide arrays, see Arrays
Wide deployment 69

MIX
Papier aus verantwortungsvollen Quellen
Paper from responsible sources
FSC® C105338

If you have any concerns about our products, you can contact us on
ProductSafety@springernature.com

In case Publisher is established outside the EU, the EU authorized representative is:
Springer Nature Customer Service Center GmbH
Europaplatz 3, 69115 Heidelberg, Germany

Printed by Libri Plureos GmbH
in Hamburg, Germany